Science and the Marketplace
in Early Modern Italy

Science and the Marketplace in Early Modern Italy

BRENDAN DOOLEY

LEXINGTON BOOKS
Lanham • Boulder • New York • Oxford

LEXINGTON BOOKS

Published in the United States of America
by Lexington Books
4720 Boston Way, Lanham, Maryland 20706

12 Hid's Copse Road
Cumnor Hill, Oxford OX2 9JJ, England

Copyright © 2001 by Lexington Books

All rights reserved. No part of this publication may be reproduced, stored in a retrieval system, or transmitted in any form or by any means, electronic, mechanical, photocopying, recording, or otherwise, without the prior permission of the publisher.

British Library Cataloguing in Publication Information Available

Library of Congress Cataloging-in-Publication Data

Dooley, Brendan Maurice, 1953–
   Science and the marketplace in early modern Italy / Brendan Dooley.
   p. cm.
   Includes bibliographical references and index.
   ISBN 0-7391-0232-X (cloth : alk. paper)
   1. Communication in science—Italy—History—17th century. 2. Science news—Italy—History—17th century. I. Title.

Q225.2.I8 D66 2001
509.45'09'032—dc21

                              00-051465

Printed in the United States of America

∞™ The paper used in this publication meets the minimum requirements of American National Standard for Information Sciences—Permanence of Paper for Printed Library Materials, ANSI/NISO Z39.48–1992.

Contents

Preface		vii
List of Abbreviations		ix
Introduction		xi
1	The Crime of Galileo	1
2	The Business of Astrology	20
3	Printing Natural Knowledge	41
4	Nature and the Universities	63
5	Teaching and Learning	87
6	Saving the Jesuit's Skin	112
7	Science and the Marketplace	138
Epilogue		161
Index		167
About the Author		179

Preface

The study of nature, Walt Whitman once inferred, is no substitute for nature itself. Nor is the history of science any substitute for the study of nature. At times, the historian may be able to convey something of the circumstances of natural knowledge, by bringing a few of its seekers, in some way, back to life. Such is the purpose of this interpretation of post-Galileian science.

The endeavor depends on research done over several years for several different projects. The theme began to emerge first in a study of eighteenth-century scientific journalism. The purview then widened to encompass the origins of the first scientific periodicals in the seventeenth century. At a still later stage this investigation converged with a study of the seventeenth-century book trade, and then, with a study of early modern cultural organizations including universities and academies. The reader will now find a synthesis that benefits to some degree from a long perspective.

At all times, I have attempted to pay as much attention to the changing forms taken by the expression of ideas about natural reality as I do to the ideas themselves. Without invoking the overwrought McLuhanesque phrase, there is much to suggest that in many cases idea and expression were indissolubly connected. How a practitioner divulged his results was often as important as the results themselves; changes in the means for divulging results had a powerful effect on intellectual processes. In the following chapters, the analysis of ideas is accompanied by the analysis of the means of their distribution and the history of ideas is combined with the history of communication. Since communication took place in a wider world of exchange, the history of communication is joined to the history of economies.

In harmony with the empirical temperament of many of the characters in this story, a good deal of the research for this study was carried out in the archives. In the State Archive of Rome I discovered the trial record of an early seventeenth-century practitioner. In the Marciana Library in Venice, I located the personal papers of several early eighteenth-century practitioners; in the Biblioteca Augusta of Perugia and the Biblioteca Civica of Padua I studied the papers of still others. In the archive of the Accademia Patavina in Padua and in the Biblioteca Bertoliana of Vicenza I found the minutes of the academies; in the Biblioteca Nazionale Centrale in Florence, I utilized the the correspondence of that indefatigable letter-writer, Antonio Magliabechi. In all these places and many more mentioned in the notes, I benefited from the friendly guidance of librarians and archivists.

And most of all, I benefited from the insights of the many friends and colleagues who over the years have contributed in some way to the ideas in this book. It is a pleasure here to acknowledge them, but I will name only a few: Allen Debus, Liah Greenfeld, Salvatore Ciriacono, Mordechai Feingold, Ugo Baldini, Piero Del Negro, Mario Infelise. All these I thank; too late, unfortunately, in the case of Marino Berengo.

Abbreviations

ASF	Florence, Archivio di Stato
ASV	Venice, Archivio di Stato
ASR	Rome, Archivio di Stato
ASV	Rome, Archivio Segreto Vaticano
BAR	Rome, Biblioteca Apostolica
BCV	Verona, Biblioteca Civica
BNM	Venice, Biblioteca Nazionale Marciana
BNC	Florence, Biblioteca Nazionale Centrale
DBI	*Dizionario biografico degli italiani*
DSB	*Dictionary of Scientific Biography*
MAP	Florence, Medici Archive Project, Documentary Sources for the Arts and Humanities
MCP	Padua, Museo Civico
PBA	Perugia, Biblioteca Augusta
QS	*Quaderni storici*
Riformatori	*Riformatori dello studio di Padova*
RSI	*Rivista storica italiana*

Dating has been modernized throughout. In transcriptions of documents, carets (^ ^) indicate marginal and interlinear insertions.

Introduction

Like other products of human ingenuity, science, too, has had its marketplace. And no one needs to be reminded about the powerful effects a marketplace has had on science in very recent times. Instruments and laboratories cost money. Even the most low-tech practitioners have to be paid. The provision of funds for these purposes requires the mobilization of resources deriving in one way or another from the other productive sectors of society. The products of science themselves, including patents and processes, are subject to mechanisms of exchange. The value of certain discoveries and certain research programs is negotiated between producer and sponsor. These features of the economic ramifications of science and their relation to commercial mechanisms in our own times are well known.[1] But what about in past times? To what extent did an emerging marketplace affect that loose collection of heterogeneous fields of natural knowledge known in the early modern period as natural philosophy, the distant cousin of modern "science"?[2] This book seeks to find out.

The investigation considers the careers of some notable practitioners and their efforts to put across their messages. It takes account of the various publics they addressed and the institutions where they conducted their activities—academies, universities and courts. It considers the new methods of diffusion they used to bring the products of their reflections to wider audiences. And it traces back to the seventeenth and eighteenth centuries some of the familiar structures of what we now call science.

In many ways, Italy seems like an ideal central focus for a study of this kind. Here the spectacular career of Galileo, pioneer in the communication of natural knowledge, was followed by a notable series of discoveries in medicine and natural history.[3] Here emerged the first academies in Europe devoted to natural knowledge—from the precocious Accademia dei Lincei in Rome to the more successful Accademia del Cimento in Florence. Here some twenty-five major universities provided an important conduit between natural philosophical and mathematical research and the non-expert. Here some of the first journals spread the latest natural knowledge to anyone who could read and follow their arguments. In each of these phases of activity, our inquiry focuses on the ways in which natural philosophers addressed their publics, showing how the exigencies of publication and diffusion influenced the science they produced.

Yet in Italy too, the emergence of a market for science received some of its most spectacular checks and encountered some of its most significant obstacles. With power over intellectual productivity contested between church and state, the opportunities for impeding the free circulation of intellectual products were nearly as numerous as the opportunities for furthering it.[4] The inquiry will thus have to cover the limits to the marketplace as well as its extent, the occasions for its circumscription as well as the occasions for its expansion.

We will observe the world of a privileged intellectual elite at the moment when Europe was being flooded with worldly goods at an unprecedented rate. Natural philosophers too began seeking the good things of life, with an alacrity that no longer seems shocking in our post-romantic age. Their studies became objects of trade with anyone inclined to reward them for what they did. Even Galileo was not immune to the general trend. And from this point of view, the members of Galileo's school and their disciples—Benedetto Castelli, Evangelista Torricelli, Michelangelo Ricci, and the rest, all of whom ended their lives in more or less comfortable circumstances—are notable examples of success. In the eighteenth-century heyday of "useful" science, the possible rewards for the study of nature became even more tempting. Applications of science to agriculture and public health increased the demands on researchers. Galvani and Spallanzani did not make huge fortunes from their discoveries. But they experienced the expanding opportunities for public sponsorship.

Science was profoundly changed as a result. From the period of Galileo onward, public spectacle began to play a larger role. Support from non-specialists became a powerful tool for drawing attention to research programs of every sort. The media of information became important instruments in this endeavor. In Italy, the needs of university students combined with political and ideological factors to determine a shift toward the life sciences and away from physics and mathematics. In the eighteenth century, natural history acquired particular importance due to the agricultural and demographic crises that invested the whole peninsula. A new emphasis on ideas with practical applications related to civic welfare provided impetus for new areas of study. That the promise of a field like electricity always remained several strides ahead of its benefits did little to dampen the public enthusiasm that fuelled specialized contributions.

Exploring the various stages in the development of the market for science will reveal the emergence of what might be called a scientific public sphere.[5] For the products of reflection in natural philosophy were obviously not just ordinary objects of exchange. Nor were they objects of contemplation by their makers alone. Like the products of political reflection, they might insinuate into the minds of the subjects of a state ideas potentially at variance with the verities guaranteed and endorsed by established authorities. At one moment, fields of thought served as repositories of notions that might be put in service by institutions belonging to the sphere of political and religious power, such as Aristotelian physics, with its suggestive hierarchical scheme of natural reality. At the next moment, such fields were being constituted as part of a sphere of public reflection in contrast to reigning orthodoxies.[6] This is the sense in which Jürgen

Habermas' model for the emergence of a public sphere might be useful for understanding the development of a scientific marketplace. In the political arena, the emergence of printed political expression, newspapers and eventually the critical journals of the early eighteenth century, contributed to the formation of an independent social space where subjects might make public use of their reason. So also in science.[7] The first genres of printed scientific publication were eventually followed by the first scientific journals. By the eighteenth century, the periodical method had been applied so successfully and discussions about natural reality became so widespread, that even the authorities were put on the defensive.

This research suggests, then, that Italian science between the age of Galileo and the age of Galvani and Volta underwent two revolutions, not one. The first concerned methods of investigation, and it has received a considerable amount of scholarly attention. The second revolution concerned methods of diffusion, and this has hardly been studied at all. Indeed, the transition to a public sphere of science was more complex and less direct than is usually believed. The purpose of this book is to show that changes in the relationship between science and the marketplace involved a process of theoretical pondering and practical tinkering that is as important as the changes themselves. The story of how the results and accomplishments of the first revolution were consolidated, matured, discussed, and made parts of the institutions of society—the universities, the academies, and even state governments—by means of the revolution in methods of diffusion, is nonetheless one of the most important episodes in European history.

Interdisciplinary in approach, the inquiry traces changes in the way natural science practitioners sought to diffuse their ideas. In these changes, it identifies clues to the emergence of the profession of university science researcher as well as to the shifting preferences of patrons. There are two underlying assumptions: first, that the scientific accomplishments of the period are part of a cultural framework connecting the study of nature to literature, philology, philosophy, metaphysics, religion, and even law; secondly, that they are best understood in social and political contexts that include academies and universities as well as princely courts. It goes without saying that modes of cultural expression in science are here studied taking account of their independence from, as well as their dependence on, socioeconomic realities.[8]

The first chapter accordingly covers the origins of the new marketplace for science in the career of Galileo. In the regime of natural science knowledge that prevailed before Galileo's time, communities of the learned exchanged information at a leisurely pace. Improvements in the efficiency of mail delivery and the formation of correspondence networks accompanying the development of territorial states made possible more frequent and efficient contacts. Competition increased between practitioners working in different parts of Europe to diffuse ideas by print as well as by mail. Galileo, by publishing his work in vernacular Italian and cultivating an informal style, sought to argue his scholarly validity claims before an audience including non-experts. Opposition to his activities centered not only on the ideas themselves but also on the ways in which he distributed them. The ensuing debate about the effects of reading had important implications for every kind of research.

A restless search for the rewards due to natural philosophical cogitation affected not only the fields of inquiry explored by Galileo, but also fields that would later be branded as "pseudo" science because of the unverifiability of their results. Chapter 2 goes into the monastery of Santa Prassede in Rome, where Orazio Morandi sought to advance his career as an astrologer in much the same way Galileo had advanced his career as an astronomer. At a time when the distinction between the two roles was still unclear, he offered the products of his observations to very much the same audiences. Like Galileo, he sought to distinguish his own products from his rivals' by a self-consciously modernizing methodology. Like Galileo, he sought to draw attention to his own mastery of these techniques by spectacular discoveries. In Morandi's case, the discovery in question was not a new star or planet or a new cosmological theory, but the interpretation of a celestial sign that spelled death for Pope Urban VIII. Unfortunately for Morandi, in announcing his interpretation, he got into far more trouble than Galileo did for announcing new proofs of the heliocentric theory. The outcome of his prolonged court case may well have determined the outcome of the Galileo case that was to occur the following year.

Where Galileo differed from Morandi was, among other things, in his use of the printing press for diffusing information about his discoveries. Indeed, as the seventeenth century wore on, a chief advantage of Galileo's followers over their adversaries was their successful implementation of strategies for spreading their ideas, as chapter 3 shows. This did not happen all at once. But after some initial resistance, Galileo's notions about a wider public sphere of science finally caught on in Italy toward the end of the century among practitioners who were interested in doing observations, performing experiments, and challenging traditional views. This new interest was exemplified in a myriad of genres of publication, including textbooks as well as broadsheets, treatises as well as pamphlets, all aiming to take the task of public instruction back from the religious orders and helping to build support for the notion of science as a socially responsible set of activities. Within the Church discussions focused on exactly which publics might be damaged by the spread of philosophical heterodoxy and in what ways.

While these discussions were going on concerning the realm of print, the diffusion of knowledge continued in another public forum: the universities. In spite of their reputation for backwardness and traditionalism, they were nonetheless places of encounter between the bearers of natural knowledge and various audiences, comprising persons in many walks of life. While governments used universities at least in part as centers of technical expertise, as chapter 4 shows, seekers after natural knowledge found in them protective havens from the tumultuous world of court patronage. The prevalence of jobs in medical faculties helped bring about a shift in Italian contributions toward biology and away from more controversial fields like mathematics and physics. And in this period the universities underwent a division of labor, whereby professors more interested in teaching the traditional texts prepared students for responding to questions on the exams. More innovative professors taught their own research, helping to form the audience for experimental science that Galileo had envisioned.

Rediscovering the cultural role of the early modern universities requires finding out exactly what was taught. However, accurate records are few and far between. Chapter 5 examines the complete lesson plans left behind by two early eighteenth-century professors at the University of Padua, Antonio Vallisneri and Giovanni Poleni, and Vallisneri's notes to the lectures of his maestro at Bologna, Giovanni Girolamo Sbaraglia. All of these professors, the documents reveal, used their lectureships to promote their own views of nature. Sbaraglia invented new visual demonstrations for conveying his eclectic Galenism. Vallisneri pioneered teaching the experimental method in biology, and we can even recreate some of the demonstrations he conducted in class. Poleni developed Vallisneri's experience-oriented lecture style into a full-blown program in physics and mechanics, winning the unqualified support of the Venetian government. Responding in part to student demand, the Venetian government thereupon ordered the construction and furnishing of a state-of-the-art laboratory for teaching and research. Experimental philosophy, at Padua, recognized as a regular part of the official curriculum, furnished an example for the acceptance of innovative ideas elsewhere.

By the second half of the eighteenth century, a public for natural knowledge of all sorts had clearly emerged. As periodical publications fed a growing demand, new opportunities opened up for using science as an enhancement to reputation. Chapter 6 records Francesco Antonio Zaccaria's appeal to a wide audience to reverse current negative judgments regarding the historic cultural role of the Jesuit order. In his Modena-based *Storia letteraria d'Italia*, Zaccaria introduced readers to the accomplishments of his contemporaries (not just Jesuits) in all the disciplines of the arts and sciences. In numerous articles devoted to science, he contributed to forming what he viewed as the modern characteristics of the scientific professions and stoked the patriotic ardor of young practitioners interested in vindicating the accomplishments of Italians vis a vis their north European counterparts. And although his effort invited almost as many expressions of skepticism as it did nods of complicity, it indicated a notable shift in Church policy away from the suppression of science to cautious support.

The creation of a wider market was not the only cause of the late great eighteenth-century scientific resurgence. But, as the seventh and final chapter shows, it may have been one of the most important ones. Perceptions of the deepening economic and social crisis of the peninsula directed attention to new ways and means for applying specialized knowledge to practical problems of public welfare. New academies of agriculture sought to combine technical interests with purely scientific ones in a way never explored in the scientific academies of the previous century. First in Florence and then in the rest of Italy, these initiatives showed the problem solving potential of science in familiar areas of experience. And to the application of science to civil life, such well-known figures as Alberto Fortis, Giovanni Targioni Tozzetti and Lazzaro Spallanzani dedicated themselves. As the language of progress and utility began to invade even the more esoteric realms of scientific discourse, a new field like electricity raised many expectations. Galvani conducted his experiments in an environment in which the study of the earth, of the animals, and of man were all conducted as part of a larger project of human cultural and social development. He viewed his own results as episodes in his ongoing efforts to find scientific ways to relieve suffering and prolong life by the most advanced methods. Although Galvani's discovery of animal electricity was eclipsed by Volta's discovery of chemical electricity, the work of both practitioners added to visions of a future ordered by technological advancement.

Late eighteenth-century visions of a technological future were accompanied by visions, at least as vivid, of a future beyond technology. The first stirrings of preromanticism were accompanied by a radical antimodernist viewpoint. Worried by the dangers of uncontrolled advancement, self-styled philosophers appealed for a return to earlier modes of knowing, brought into the current cultural milieu by way of Mesmerism, Galvanism, animism, dowsing and even the evil eye. While the period brought in many of the tools that would build a public discussion of the uses of knowledge, it also introduced many of the perplexities that would remain the hallmarks of modernity. For practitioners of all the natural sciences, these perplexities constituted a new and powerful challenge to their day-to-day activities. Whether they liked it or not, henceforth they depended to a considerable extent on the public interest they were able to stir up for their productions.

The relation between science and the marketplace has not always been an easy one. But it has always been productive—if not of the most advanced ideas (however we might wish to define those), at least, of intellectual energy and ferment of the most varied sort.

Notes

1. Consider *Europe-Asia—Science and Technology for Their Future: Science, Economy, Culture*, 7th conference, 26-28 March 1996, Forum Engelberg (Zurich: ETH Zurich, 1996); Charles Edquist, ed., *Systems of Innovation: Technologies, Institutions, and Organizations* (New York: Pinter, 1997); H. Radder, *The Material Realization of Science. A Philosophical View on the Experimental Natural Sciences Developed in Conjunction with Habermas* (Assen: Van Gorcum, 1988).

2. Patronage studies are the best-developed field in this regard. In addition to the more specific studies mentioned in subsequent chapters, Bruce T. Moran, ed., *Patronage and Institutions: Science, Technology, and Medicine at the European Court, 1500-1750* (Rochester: Boydell, 1991).

3. Paolo Casini, *Introduzione all'Illuminismo*, 3rd edition, 2 vols. (Bari: Laterza, 1980), vol. 1: *Scienza, miscredenza, e politica*, 222; Cesare S. Maffioli, *Out of Galileo. The Science of Waters, 1628-1718* (Rotterdam: Erasmus, 1994); Ugo Baldini, "La scuola galileiana" and "L'attività scientifica del primo Settecento," in *Storia d'Italia*, Annali 3: *Scienza e tecnica* (Turin: Einaudi, 1980), 383-551. Compare Michael Segre, *In the Wake of Galileo* (New Brunswick, 1991).

4. Mario Infelise, *I libri proibiti, da Gutenberg all'Encyclopedie* (Bari: Laterza, 1999). Compare Roger Shattuck, *Forbidden Knowledge, from Prometheus to Pornography* (New York: St. Martin's, 1996).

5. Compare T. Broman, "The Habermasian Public Sphere and Science in the Enlightenment," *History of Science* 36 (1998): 123-49.

6. For reflections on this contrast, Vincenzo Ferrone, *The Intellectual Roots of the Italian Enlightenment: Newtonian Science, Religion, and Politics in the Early Eighteenth Century*, trans. Sue Brotherton (Atlantic Highlands, N.J.: Humanities Press, 1995) [orig. ed., *Scienza natura religione: mondo newtoniano e cultura italiana nel primo Settecento* (Naples: Jovene, 1982)].

7. Jürgen Habermas, *The Structural Transformation of the Public Sphere: An Inquiry into a Category of Bourgeois Society*, trans. Thomas Burger (Cambridge, Mass.: MIT Press, 1996); concerning the role of print in science, Elizabeth Eisenstein, *The Printing Press as an Agent of Change* 2 vols. (Cambridge: Cambridge University Press, 1979), vol. 2; William D. Garvey, *Communication: The Essence of Science* (New York: Pergamon Press, 1979) A. J. Meadows, ed., *The Development of Scientific Publishing in Europe* (Amsterdam: Elsevier Scientific Publishers, 1980).

8. I thus favor the interpretation of H. Floris Cohen, *The Scientific Revolution: A Historiographical Inquiry* (Chicago: University of Chicago Press, 1994) over that of, say, François Russo, *Libres propos sur l'histoire des sciences* (Paris: Blanchard, 1995), 10, 39. Concerning this problem, I recommend Ian Hacking, *The Social Construction of What?* (Cambridge, Mass.: Harvard University Press, 1999); as well as the debate surrounding Paul R. Gross and Norman Levitt, *Higher Superstitution. The Academic Left and its Quarrels with Science* (Baltimore: Johns Hopkins, 1994). Consider Steve Fuller, "A Tale of Two Cultures and Other Higher Supersitions," *History of the Human Sciences* 8 (1995): 115-25, and the review by Norton Wise in *Isis* 87 (1996): 323-27, as well as the reply by Gross and Levitt in *History of the Human Sciences* 8 (1995): 125-29.

Chapter 1:
The Crime of Galileo

When did the notion emerge that knowledge about the natural environment ought to go public rather than be kept as the special preserve of a few adepts? Perhaps at various times across the centuries. But never was it so hotly contested as in the seventeenth century; by no one was it so powerfully supported as by Galileo Galilei. After he joined the hitherto distinct genres of scientific prose and vernacular rhetoric in a powerful, new, and persuasive synthesis and sent his works to the printer for distribution, the communication of natural knowledge would never be the same again. The dramatic changes that occurred between the sixteenth and the seventeenth centuries in the disciplines loosely grouped under the label "natural philosophy" and later to be known as "science" seem to be summed up in his career. Let us place this now somewhat neglected aspect of the "Galileo Affair" into clearer focus.[1]

Within the context of late Renaissance theories of reading, the dangers that ecclesiastical and university authorities perceived in Galileo's effort to diffuse his discoveries make much more sense, and his innovations seem all the more strikingly original.[2] Among the many readings of the "Affair," one that takes account of the role of the marketplace for natural knowledge seems to answer more of the lingering questions and promises to provide a more meaningful basis for our investigations of the period that followed.

It might be helpful to recall the major issues.[3] In 1616, partly in response to writings by Galileo, Nicholas Copernicus' *On the Revolutions of the Heavenly Spheres*, written over half a century before to prove that the earth moved about the sun, was prohibited "until corrected." The implication of this action was that heliocentrism was a dangerous topic; and Galileo received a friendly warning to this effect from none other than Cardinal Robert Bellarmine. It was an unusual intrusion of the Church into what seemed to some observers (including Galileo) like purely intellectual matters out of its purview. Believing that the truth should and would eventually prevail, he went on to complete the *Dialogue Concerning the Two Chief World Systems*, a heliocentric tour de force. He anticipated some trouble from the prepublication censorship mechanism that was currently used in all the states of Italy as well as everywhere else in early modern Europe. To stay on the safe side, he had the work approved not by just any censor but by the

1

Master of the Sacred Palace in Rome, and later by the Florentine Inquisitor so it could be printed in Florence. None of this sufficed; and he was condemned for his views as well as for paying insufficient attention to the warning of Bellarmine, who was now dead and unable to defend him.

One problem of interpreting the events is that the evidence itself is fraught with ambiguities. To cite just one example, Galileo carried with him to the trial an affidavit signed by Bellarmine in 1616 saying that there had been no personal censure against him in the Copernicus condemnation and repeating the friendly warning, given in person at their meeting, to steer clear of such matters. If this is all that happened at the meeting, Galileo could be expected to publish the *Dialogue*, as long as he was careful. The officials at the 1633 trial produced a transcript of the Bellarmine meeting in which Galileo was specifically enjoined not to write or say anything concerning heliocentrism. In this case, the *Dialogue* was a foolhardy violation. One theory, still unverified, is that the transcript was scribbled into the register by Galileo's enemies after the Bellarmine meeting, to ensure his downfall.

Over the years, the events have been scrutinized from two basic standpoints. One approach, let us call it "epistemological," has examined the contents of the debates, attempting to trace the emergence of a new cosmology and the faith-science controversy in the history of thought. Another approach, let us call it "sociological," has focussed on circumstances external to the debates themselves that may have determined the outcome—the personalities of the protagonists, the private grudges, the public interests, the structure of authority. Recent scholarship, attempting to combine these two approaches, seems to have raised more questions than it has answered.

In one version, a cryptographic reading of certain hidden innuendoes in the documents, combined with a microscopic view of the cultural environment, reveals that the controversy was not about heliocentrism at all.[4] Instead, Galileo's misfortunes derived from his support for the matter theory known as atomism. By contending that the outward appearances and properties of things came from the various combinations of tiny particles that made them up, he put himself in the center of what was to become, after heliocentrism, the second noisiest debate of seventeenth-century science. And for the Church, the stakes in this debate were almost as high as they were in the heliocentric debate. If atomism took root, what would happen to views about the Eucharist during Transubstantiation, based on the scholastic form-plus-substance analysis of matter? The transformation of the bread and wine into the body and blood of Christ seemed somehow comprehensible in terms of a change of substance with unchanged form. If the appearances of things were produced not by form but by various combinations of tiny particles, how could the body of Christ seem like bread? No wonder Church officials ganged up on Galileo. And their campaign against the *Dialogue* was merely a convenient pretext for an attack on *The Assayer*, where Galileo gave the bare outlines of a theory of matter based on tiny particles.

But perhaps, as another version suggests, Galileo's downfall was not due mainly to his ideas at all. Perhaps it was due to his involvement with the structure of baroque authority known as the "court."[5] Surely, the key to advancement in seventeenth-century society was successful courtiership, and Rome was the court par excellence. Moreover, from the point of view of the ruler's prestige, executing swift and unmitigated justice on a turncoat courtier was as important as having good courtiers around. Galileo, having gained renown by his courtiership in Florence, experienced a precipitous fall from grace as a client of Pope Urban VIII. Although not officially a papal courtier, by taking his own arguments and himself too seriously in the *Dialogue*, this interpretation goes, Galileo showed himself to be a potential nuisance and a bore. And just then, the pope was in need of eliminating all potential nuisances and bores around him who might detract from his prestige and interfere with his task of facing down the French and the Spanish in the Thirty Years' War. The trial was merely the formal mechanism for bringing about this predetermined outcome.

There is some truth in all of these views. No doubt, the Galileo Affair belongs to the history of ideas just as much as it belongs to the history of society or politics. No doubt Galileo's ideas on many subjects collided with contemporary conceptions, and not just those concerning the movement of the earth. And no doubt networking was as helpful for advancement in seventeenth-century society as disappointing one's patrons was dangerous.

But even if we look at events strictly from the perspective of the history of ideas, Galileo's crime lay less in what he said than in how he said it. After all, few of the ideas in the case originated with him; some, for instance, came from Copernicus.[6] And many contemporaries who held similar ones were not persecuted—for instance, the French philosopher-cleric Pierre Gassendi and his fellow countryman, the mathematician Gilles Personne de Roberval. Nor was the crime directly concerned with his patronage arrangement. From the outset of his career, Galileo deliberately sought to transcend the narrow boundaries of courts and to secure his position by gaining a broad base of support for his ideas outside among the general public.

In his now-classic book, *The Crime of Galileo*, Giorgio de Santillana suggested that Galileo's crime lay in his effort to appeal over the heads of contemporary authorities by taking his case for a new scientific approach and a new cosmology directly to the people. There were significant problems with the way Santillana presented his thesis. To some degree his argument was flawed by his pressing desire to show parallels between the Galileo case and the case of J. Robert Oppenheimer in his own time. In both cases, he argued, a respectable institution (the Church on the one hand, the U.S. government on the other) had been badly served by the misguided zeal of its own bureaucrats and the benign neglect of its leaders. However, unlike in the 1950s, there was no witch hunt in the first decades of the seventeenth century.[7] These were decades of relative relaxation in the activities of the Inquisition, at least compared to earlier decades. Furthermore, Galileo's adversaries directed their anger almost exclusively at Galileo himself. And the case of the philosopher Giordano Bruno from Campania was not, as Santillana suggests, of the same sort; although the documents have been lost, he was burned in 1600 most probably for his views about the transmigration of souls, an almost exclusively religious question, and not for his astronomical ideas.

Nonetheless, a better understanding of the novelty of the way Galileo communicated his scientific ideas suggests that Santillana's thesis may be well worth modifying and presenting anew. And to appreciate this novelty, it is helpful to consider first of all what scientific communication was like before Galileo began his campaign. Then his behavior can be seen in the light of what, extending Santillana's thesis a little, might be considered the most important contemporary influence of all—namely, the emergence of new practices in the communication of natural knowledge. In the period when Galileo began his career, going public with one's views was still unusual, and as it turned out, could even be a crime.

Before the time of Galileo, communication of original results to a public of experts was not viewed as essential to natural philosophy. Important ideas were transmitted at least as frequently through particular traditions and particular schools as through letters—such as the Bolognese school of mathematics and the Paduan school of philosophy. Reputations of those researchers who had academic positions were sustained at least as effectively by public disputations as by publication.[8] An example was Filippo Fantoni, Galileo's mid-sixteenth-century predecessor as lecturer in mathematics at Pisa, who published one treatise on the reform of the calendar and then left all the rest of his works in manuscript.[9] Intellectual property was protected mainly by secrecy. Examples abound, but here are two of the most famous in Italy. The sixteenth-century Brescian mathematician Niccolò Tartaglia established his scientific career by solving the cubic equation independently of the solution's first discoverer, Scipione dal Ferro at Bologna, and by using it in an academic duel with the one student to whom dal Ferro had confided it: Antonio Maria Fior. Instead of publishing the solution, Tartaglia also kept it to himself until he succumbed to requests for it by Girolamo Cardano. Slightly later, Francesco Maurolico published only part of an optical work, and not even the best part; what was most innovative he circulated in manuscript copies among his students in Sicily.[10]

None of the various forms of scientific publication were designed to exploit all the advantages of print. Important treatises were published, almost as an extension of person-to-person communications, for a tiny public of experts and with no particular urgency. Copernicus' *On the Revolutions of the Heavenly Spheres*, published over two decades after the research had been completed and in very special circumstances regarding the Roman authorities, may be an extreme example. But our knowledge about Copernicus suggests his aim was mainly to get just a few others interested in what he believed were uncompromisingly abstruse problems, and in no particular hurry.[11] Copernicus's contemporaries evidently preferred bulk to speed, and they saw printed communications simply as extensions of the nonprint ones that really counted. Shorter communications, such as reports on observations, they filled out with elaborate appeals to the reader, frequently consisting of references to personal meetings. Tartaglia's *Nova scientia*, concerning a new way of measuring the paths of projectiles, opened with a long description of a meeting with an eminent artilleryman.[12] Whenever possible, they collected short communications into single more imposing volumes rather than publishing them right away. By this approach, no doubt Leonard Fuchs hoped to make the same impact in Tübingen by his collected *Libri quattuor* as Leonardo Fioravanti hoped for in Bologna from his *Treasure of Human Life*.[13]

The initial impeti to the formation of natural history collections are as complex as the collections themselves were many-sided. But at least one motive, judging from the career of Ulisse Aldrovandi, may well have been in order to overcome some of the disadvantages of preprint communications.[14] Aldrovandi's ambitious effort to establish the relationships between all the fields in the encyclopedia of human knowledge called for the the collaboration of correspondents from all over Europe. More than scientific views or news of recent publications, we find his correspondence to consist largely of requests for specimens of natural objects that he had been unable to collect on his own numerous voyages. If the leisurely pace of public postal delivery made the exchange of observations agonizingly slow, Aldrovandi could nonetheless rely on his own personal engagement with the evidence. After thirty years he was able to claim, "I have collected in my theater of nature more than thirteen thousand things of every kind, inanimate and animate."[15] And only when he had collected everything could he be sure to surpass in originality the conclusions of contemporary researchers he was unable to consult regularly. "There is not a single thing under the moon that cannot be reduced to one or another of the categories of inanimate things or fossils dug out of the earth, plants, or animals," he noted, "of which [my scribes and I] have written the detailed histories."[16] When he requested funding for the employment of scribes, it was not in order to improve his epistolary communications with other scholars but to make sure that none of the interpretations pouring forth from his own head would fail to be written down. And to make sure everyone knew about what he was doing, it was enough for him that his handiwork was seen by "all the gentlemen who come to town, who visit my *pandechia* as the eighth wonder of the world."[17]

Attempts were made, in the sixteenth century, to communicate natural knowledge to a wide audience.[18] Leon Battista Alberti's patriarchs in *On the Family* and Baldassarre Castiglione's courtiers in *The Courtier* continued to serve as models for a well-rounded education. And a well-rounded education was thought to include some elementary information about cosmology and mathematics.[19] At a lower social level, specialized knowledge about the properties of matter and the mechanics of solids had long been viewed as a necessary accompaniment to the other skills of architects and shipwrights. No one forgot that Daniele Barbaro had translated Vitruvius' *On Architecture* into "the words of our Arsenal" precisely in order to "help encourage practice in these matters."[20]

A literature on natural knowledge in the spoken language instead of in Latin was supposed to inspire new generations of practitioners in Italy just as vernacular science had done in ancient Greece and Rome. For too long, noted Alessandro Piccolomini, the study of the sciences had been limited to grammar-school students with little time for creative thought. "If [students] want to know something," he observed, "they find it necessary to consume not only their childhood and adolescence—their most impressionable age—but even their youth and a good part of their maturity, which are the most powerful times for our intellect, in accents, words, and structures."[21] Yet there was no proof that grammar students were potentially better philosophers than anyone else, and those of them who finally finished the course were already so worn out that they were unlikely to come up with anything original. If, on the other hand, students could learn the "sciences themselves" in the vernacular like the Greeks and Romans, without the interference of a foreign language, Piccolomini suggested, new Italian Platos, Aristotles, Ptolemies, and Galens were sure to spring up spontaneously.[22] Sperone Speroni agreed. "If . . . future philosophers reason and write about science [in Italian]," he insisted, "the intellect and sentiment [of the sciences] will be the same as that of amateurs and of those who study the doctrines buried not in languages but in the hearts of men."[23] That, he believed, would mean a definite step forward.

Toward the end of the sixteenth century, improved communications threatened to render previous ideas about the publicizing of research results obsolete. Within states, the first regular biweekly, weekly, fortnightly, and monthly statewide letter services replaced the previous system of haphazard couriers and messengers. Between states, governments began to set up public letter services alongside official diplomatic ones to spread the benefits of dependable mail beyond a small circle of merchants, diplomats, and university dons and guaranteed them by treaties with surrounding states. In 1608, Ottavio Codogno codified all the new developments in his manual, *New Itinerary of the the Mail in the Entire World*.[24] The breakdown of geographical barriers helped spread information about innovative ideas over wider and wider areas. Great letter systems, those of French scholars Nicolas Claude Fabri de Peiresc and Marin Mersenne, replaced occasional correspondence as the main means of communication.[25] The latter, by his keen eye for scientific excellence, managed to turn his monastery in Paris into a veritable clearinghouse of information, by putting scholars in contact with one another at opposite ends of Europe, from Brussels to Florence to Oxford.

The Roman philanthropist Federico Cesi founded the Accademia dei Lincei to ensure that the changes in communication occurred in an organized fashion. He called for improving correspondence by a new standard format for scientific letters. More importantly, he encouraged researchers to publish, since containing major scientific events among a small circle of experts was evidently impossible. "Anyone can imagine," he exclaimed, "how much honor and esteem [an author] can gain by such a communication, and how much he can make himself known both to princes and other men of letters and to the whole populace."[26] And around the same time that Francis Bacon was making similar proposals in England, he called upon all researchers to establish connections with their colleagues and collaborate, on an unprecedented scale, in a project for realizing all the possibilities offered by the latest technological inventions.

These changes occurred in a crucial moment from the standpoint of the observation of nature. For it was just at this time that Johannes Kepler began to understand the importance, for other problems besides pure algebra, of John Napier's newly invented technique of logarithms. Calculations of stellar angles that once took several days or even weeks to complete could now be carried out in a matter of minutes by consulting a few handy tables.[27] And when Galileo and Thomas Herriott, at opposite ends of Europe, first gazed at the heavenly bodies through the lenses of a new optical instrument invented for magnifying distant objects, the age of telescopic astronomy began—with enormous consequences. More and more astronomers thought to try their hands at manipulating the new instrument. And while the number of disputes multiplied about the priority of this or that discovery of this or that celestial appearance, so also did the number of observations.

Improved communications turned out to be a mixed blessing at first. Galileo's fellow-researchers converged so closely on a number of similar astronomical problems that similar results were almost inevitable. The absence of rules on intellectual property combined with the high stakes of scientific patronage to make priority disputes both inescapable and bitter.[28] Galileo's main telescopic discoveries—the "Medicean planets" (four moons of Jupiter) and sunspots—accordingly became the targets of Simon Mayer and Christopher Scheiner in Germany. At the same time, improved communications encouraged at least one Italian colleague, Giovanni Antonio Magini at the University of Bologna, to add a new technique to university advancement: the recruitment of transalpine reinforcements. "Besides writing to [Johann Entel Zugmesser] of Cologne to put him against you," a correspondent informed Galileo, "he has done the same with all the mathematicians of Germany, France, Flanders, Poland, England, and so forth."[29] The story of Galileo's strategies for defending his ideas between 1606 and 1632, apart from the purely intellectual issues involved, belongs to the history of science and the marketplace.

Galileo quickly learned to operate in the new world of printed knowledge. After publishing the *Starry Messenger* in Latin for a relatively small scholarly audience, he immediately set aside all previous distinctions between scholarly communication and communication to the general public. Everything would now appear in pure Tuscan Italian.[30] With a fervor that his sixteenth-century predecessors had reserved for inculcating the ideas essential for solving the urgent cultural and political problems of the time, he set out to convert everyone, and especially Italians, not just to a general appreciation of the benefits of knowledge but to his own particular views—even when these had no obvious practical consequence.

Debates that began in Latin outside of Italy Galileo translated into the vernacular. In his responses to Scheiner and Mayer, he provided vernacular versions of sixteenth-century-style point-by-point academic refutations, showing that his adversaries could not have seen the things they claimed and that their explanations were faulty. In both cases, he reinforced his argument by including copies of the works impugned—in the original Latin so busy citizens could skip them and read only his own lampoons and descriptions of his own unorthodox cosmos.[31]

Debates that began as purely academic occasions Galileo finished off by opening them up to the general public. An example was the dispute between Galileo and the Aristotelians concerning why bodies float in water. When his colleagues in Florence proposed to solve the matter in good sixteenth-century fashion by a public demonstration before a clergyman, Galileo failed to show up. Instead, he put together the *Discourse on Floating Bodies*, an effective polemical tract rejecting the Aristotelian causal explanation that bodies contain air that tends upwards toward its proper airy sphere, in favor of the one of Archimedes, that bodies float because of the difference between their specific weight and water's. Considering the "novelty of the various propositions" and the possibility that "excessive vehemence and excessive raising of the voice" could confuse matters in a public dispute, "it is much easier to explain by pen than by voice." This approach allowed him to deal on his own terms with kinks in his argument concerning the role of shape in determining buoyancy; it forced his interlocutors to defend themselves against a polemicist more able than they; and it permitted him to address a larger public. "Most of [the book] ought to be understood," he claimed, "by the whole city." Accordingly, he permitted his printer to ensure that this public extended far beyond municipal boundaries—to "the many who from Venice, from Rome and from other places . . . have requested the present treatise."[32]

To be sure, there were other aspects of Galileo's behavior besides his appeal to the public that were not calculated to win him the love of his adversaries. He had a way of formulating his contributions to debates in a fashion meant to demolish the opponent's personal as well as intellectual credibility. When Simon Mayer appropriated his discovery of a surveying device called a geometrical compass, a kind of sector marked off for determining any area bounded by straight lines and curves, Galileo shot back with a *Defense against the Calumnies and Impostures*. . . . Receiving a put-down from Galileo could be a memorable experience. Of Orazio Grassi, his anonymous Jesuit antagonist in the controversy over the comets of 1618, he said:

> Those who wear disguises are either low-class persons who desire to have the esteem due only to lords or gentlemen in order to use the honor that nobility brings for their own ends; or else they are gentlemen who, surrendering the respectable dignity appropriate to their station, allow themselves, according to the custom of many cities of Italy, to talk freely about everything with everyone, taking equal pleasure from allowing any person, whoever he may be, to talk and dispute with them without any fear. And I believe that the wearer of this disguise of Lothario Sarsi is among the latter (since if he were among the former, he could derive little pleasure from passing this work off as being greater than it is).[33]

In the same book (*The Assayer*), he did not scruple to support a questionable theory—in this case, the notion that the comets were mere atmospheric and not astronomical phenomena, a position which he abandoned in later work—if it suited his immediate purposes of demolishing an argument. When his adversaries had ecclesiastical connections, such behavior meant trouble for Galileo.

Characteristically, Galileo brought to his Italian readers the most complicated question of all in what contemporary rhetorical ideas seemed to guarantee was an easily understandable form: an Italian *Dialogue Concerning the Two Chief World Systems*. Such a "philosophical comedy," he claimed, would have enough "breadth" to allow "the possibility of including" many novelties "without difficulty or affectation."[34] And to make sure that he reached both halves of his intended audience—the experts and the general public—he made his preface ambiguous enough to permit different readings by each. To the general public, he presented a "sincere" profession of conformity to Paul V's 1616 decree prohibiting the circulation of Copernicus' *On the Revolutions of the Heavenly Spheres* until "corrections" were made in its support for heliocentrism. "Some have said that the decree was the result of . . . uninformed passion," he remarked, "and . . .

of consultants who knew nothing about astronomical observations. . . ." He hoped to show, on the contrary, that "more is known about these matters in Italy and particularly in Rome" than anywhere else—and to drive the point home, he signed himself as a member of the Roman Accademia dei Lincei.[35] If these readers managed to get through some of his arguments, he hoped they might be persuaded that a diurnally and annually moving earth was not as absurd as it seemed. To his other readers, those who were already familiar with the work of Copernicus, he offered the same passages in his introduction as an exquisitely ironical critique of the Roman authorities to accompany the dialogue's brilliant parody of Jesuit Aristotelianism in the aptly-named character of Simplicio.

In undertaking this vernacular campaign, Galileo echoed the ideas of communication prevalent in the previous century. Natural knowledge, he agreed, was an essential part of education. "The glory of God is evident in all of his works," he said, "and it can be read divinely in the open book of the skies."[36] Indeed, mathematics was capable of lofting its possessor to undreamed-of heights: "Even though the divine mind knows infinitely more [mathematical] propositions, I believe [human] knowledge equals the divine in objective certainty."[37] It was not fair that those who were not learned in Latin should be cut off. "In the same way that many who apply themselves to the professions [of medicine, philosophy, etc] are totally inept," he noted, "others, who would otherwise be capable, are busy with family matters or other occupations that have nothing to do with letters."[38] For them, quick vernacular introductions were indispensable. Finally, the diffusion of science would benefit science itself. "It is a good idea for the prince to have his philosophers disagreeing and [belonging to] different sects," he wrote in an unpublished draft, "the better to find the truth."[39] And with scientific progress came direct civic benefits. "Knowing the truth," Galileo remarked, "can be useful."[40]

Existing communications theories, however, did not say anything about what Galileo wanted to do—namely, to diffuse such deliberately iconoclastic ideas as heliocentrism and anti-Aristotelian quantitative physics. He knew the danger. "The schools would lose too much attendance," he remarked, "if [students] deviated from the smooth and secure path of peripatetic philosophy."[41] The danger, however, was far outweighed by the benefits. The most advanced conclusions were bound to raise the level of popular discussions on practically every subject, since they helped improve the mind in direct proportion to their distance from popular belief. "I want [my readers] to see that nature, which has given them eyes to see its works," he insisted, "has also given them a brain to grasp and understand."[42] And most important of all, popular support for science could deter the Church and the governments from making a terrible mistake. "Prohibiting all of science would be the same as repudiating a hundred scriptural passages," he remarked to the Tuscan grand duchess.[43]

And the way to persuade this broad audience was by skillful rhetoric. Now, rhetoric had a special place in late Renaissance education, and by the time of Galileo it was regarded as having almost mystical powers to transfix the minds of readers and win them over to the writer's point of view. Consider these comments of seventeenth-century theorist Sforza Pallavicino, who, not incidentally, was also a member of the curia:

If men could manifest their ideas immediately like the angels, words would be superfluous. But since in order to understand one another it is necessary to paint those ideas with some perceptible colors, why choose the sordid dinginess of coal rather than the more gracious tints of ultramarine? Since some sort of recipient is necessary to transport this liquid from one mind to the other, what good is it if that most salubrious liquid, namely, the teachings of wisdom, is offered for drink in a filthy and stinking bowl that nauseates, rather than in a golden cup, sweet-smelling, that invites others to place their lips upon it?[44]

With the proper tastes and textures, the ideas of the writer could be poured directly into the minds of the readers.

The dialogue form chosen by Galileo was regarded as particularly effective for this purpose. For the dialogue was based, like drama, on the imitation of actual actions. "No animal loves imitation more than man," said Pallavicino, "he delights in seeing it, delights in doing it. This for the most part is the origin of the enjoyment of poetry, painting, sculpture and music; and this is the reason why man learns all the arts so easily."[45] Therefore, putting ideas across in a dialogue form was almost guaranteed to make converts to them. "Whoever imitates teaches, and whoever sees imitation learns."

Galileo's decision quite literally to go public with his discoveries was, according to his opponents within the Church, a crime. Educated in the same writing skills as he was, admittedly to less effect, they fully believed readers would take after his faultless prose with the same gusto whereby religious nonconformists had taken after outlawed Protestant tracts in the previous century, and vernacular bibles in the present one. "The job of the teacher ... is to transmit knowledge as quickly and as effectively as possible in order to produce willing and docile disciples," noted one counsellor to the Holy office, "particularly by proposing new ideas, which admirably attract curious minds. An examination of the whole [*Dialogue*] shows how cleverly and adroitly Galileo surpasses everyone in doing this."[46] No one seemed likely to come away unaffected. "He explains his doctrine so that many will be persuaded—even those who know something about mathematics."[47]

Once the damage had been done, it was thought impossible to undo. The difficulty of standing in the way of the elegant and flowing phrases of a Galileo had been carefully explained, in the context not of scientific but of political works, by Galileo's friend, the Venetian Servite, Paolo Sarpi: "never attempt to respond to writings that speak evil with brevity and wit, even if falsely, when the defense requires a long narrative or discourse, since brief and witty expressions impress themselves on and take over the mind, whereas a long discourse tires it to such an extent that it will never open up to the truth."[48] Worse yet, "not so much in private and military disputes as in literary ones, there is no worse misery than to be always on the defensive; and whoever adopts that strategy must lose, because the enemy can always be sure to be respected and not attacked."[49] Without a means of self-defense and without Galileo's gift of language, the adversaries seemed doomed.

Church officials had no doubts about the dire consequences of insinuating new and disturbing ideas into the minds of innocent seventeenth-century readers from all walks of life. "Some fantastic doctrine could be introduced into the world," Pope Urban VIII's nephew Francesco Barberini was reported to have said, "particularly in Florence, where minds are very subtle and curious."[50] Sforza Pallavicino agreed. "Recently, sometimes the purpose of the dialogue has been to show the weakness of the conventional arguments and the obscurity of the problems that the common people enthusiastically regard as self-evident, exciting in the readers the curiosity and greed of cleverly speculating."[51] No one knew where such speculation might go, but anyone could guess: against the Church, against political authority, against social customs.

Were they right? Did readers really stop to consider why the earth should be considered less perfect than the heavens or why circular motion should be considered more perfect than linear motion? Did they compare Aristotle's best case for the movement of the heavens around the stationary earth to Galileo's reports on his own observations of cosmic imperfections like spots on the sun and craters on the moon? Did they wonder at the incredible velocity whereby bodies would have to hurtle to circle the earth each day even in the special perfect celestial environment of the heavens? Galileo himself offered the utterly sensible rebuttal: "who can believe that nature . . . chose to move an immense number of huge bodies at an amazing velocity to accomplish what it could do by a simple movement of one such body around is own center?"[52] No one at the time would have had to forgive his mistake about the tides.

As things turned out, fellow astronomers and philosophers working in the same fields as Galileo were more than willing to make the sacrifices necessary to stay up to date with his latest conclusions, even when, for non-Italians, this meant procuring their own translators. And many of them forthwith abandoned geocentrism. Galileo's Calabrian acquaintance Tommaso Campanella spoke with enthusiasm of Galileo's work, even though he may never have actually converted to heliocentrism. "Everything pleased me," he wrote; "and I see how much more persuasive your argument is than Copernicus', although the latter is still fundamental."[53]

However, most readers from "the whole city," as Galileo called his general audience, proved both Galileo and his adversaries to be equally in error about the sovereign power of his rhetoric. Granted, they had few chances before 1616 to become acquainted with heliocentrism. One exception was the *The Ash Wednesday Supper* of Giordano Bruno. Supposing that only gifted initiates could understand science, he yielded up his conclusions only as a reward for considerable effort, and he called Copernicus an idiot. More "reader-friendly" was Antonio Foscarini's Copernican *Letter on the Work of the Pythagoreans*, but this work was mentioned in the 1616 decree against Copernicus and had little time to circulate. For understanding Galileo's *Dialogue*, most readers did not possess a sufficient background, even though Galileo invented the inquisitive character of Salviati deliberately to represent the intelligent layman. "To defend something so contrary to the intelligence and capacity of men," a friend warned Galileo, is almost impossible, "since there are very few who know the meaning of the observation of signs and celestial changes."[54] Even those of them who appreciated new ideas found heliocentrism too discordant with their religious convictions. A well-known example of unpersuasion can be taken from outside Italy: John Milton, who nonetheless included the 1633 condemnation in the *Areopagitica* as an example of papist obscurantism.

Worse yet, most readers were unaccustomed to devote to books the kind of attention required for following an intricate scientific argument. Instead of "opening [such books] anywhere and starting to read," noted Alessandro Piccolomini, they had to begin "from the beginning, and [follow] step by step, according to the way in which the matters depend one upon the other."[55] They were likely to find arguments in the *Dialogue* particularly difficult—as Galileo himself admitted. "I will take great pains to make myself understood," he made Salviati say, "but the difficulty of the phenomenon and the great mental abstraction required to understand it overwhelm me."[56] The number of them likely to keep alert from beginning to end could only be small. "Since the vulgar do not understand things that are written," Galileo lamented, "it is enough for them to see characters [on the page] and they believe [a given doctrine] has been refuted."[57] Observers and friends agreed. "[Everyone] is content," regretted one, "to follow a great master [such as Aristotle] like a torrent or like some other prime mover."[58]

News about Galileo's work spread unevenly, to be sure, since it was often completely ignored; but what was transmitted most widely had nothing to do with his hopes or the Church's fears. Reactions to his prepublication announcements in 1630 were typical. Consider this from one of the regular manuscript newsletters that brought political information and gossip to elite readers all over Rome: "The famous mathematician and astrologer, is in town, trying to print a book in which he opposes many opinions sustained by the Jesuits." It then went on to confuse the Galileo story with another story, which we will discuss in the next chapter, attributing to Galileo certain prophecies about Urban VIII and his family that had actually been made by others.[59]

As often happened, the most powerful force for introducing heliocentrism, at least in the 1630s, turned out to be Galileo's trial. After all, there were not too many copies of the *Dialogue* left for distribution all over Italy after Galileo made gifts to all the powerful ecclesiastical and political figures he could think of and ordered his disciples to do the same. As soon as the trial was reported in the news, even before the other researchers heard about it by letter, a few rebellious readers forthwith declared their conversion to heliocentrism. Apparently, scientific conclusions could be taken as much on faith as the truth of the Scriptures. Galileo follower Donato Rossetti commented concerning an acquaintance in Turin, "He hasn't read any modern books, but he talks about Galileo and his abjuration and says he is inclined to the Copernican system."[60] Almost everyone else no doubt emulated the total indifference of Modenese poet Alessandro Tassoni, who mentioned the dispute and sided with the Church only because he thought that would make his new collection of essays "most curious."[61]

Galileo's decision to go public was not, of course, his only crime as far as the ecclesiastical officials were concerned. But because of their basic agreement with him about the sovereign power of the written word, it was one of the most important. Nor were Galileo and his disciples alone in their experimentation with new ways of engaging the public in their campaigns to gain credit for their inter-

pretations of nature. While they and their adversaries discussed the new aims and techniques of marketing science, magicians and mystics, too, joined in, as the next chapter will show.

Notes

1. Concerning the most recent controversies about the Galileo case, Rivka Feldhay, *Galileo and the Church : Political Inquisition or Critical Dialogue?* (Cambridge: Cambridge University Press, 1995); Egidio Festa, *L'erreur de Galilee* (Paris: Editions Austral, 1995); Paul Poupard, ed., *Après Galilee: science et foi: nouveau dialogue* (Paris: Desclee de Brouwer, 1994); Dominique Tassot, *La Bible au risque de la science: de Galilee au P. Lagrange* (Paris: de Guibert, 1997); Thomas Schirrmacher, *Galilei-Legenden: und andere Beitrage zu Schopfungsforschung. Evolutionskritik und Chronologie der Kulturgeschichte 1979-1994* (Bonn: Verlag fur Kultur und Wissenschaft, 1995).

2. Helpful bibliographical information concerning the debate on secrecy and openness is in William Eamon, "From the Secrets of Nature to Public Knowledge," in David C. Lindberg and Robert S. Westman, eds., *Reappraisals of the Scientific Revolution* (Cambridge: Cambridge University Press, 1990), 333-67.

3. Apart from the other works mentioned below, the following account relies on: Ludovico Geymonat, *Galileo Galilei*, (Turin: Einaudi, 1957; Maria Luisa Altieri Biagi, *Galileo Galilei e la terminologia tecnico-scientifica* (Florence: Olschki, 1965); Alexandre Koyré, *Études Galiléennes* (Paris: Hermann, 1966); Stillman Drake, *Galileo at Work* (Chicago: University of Chicago Press, 1978), keeping in mind the review by William R. Shea in *Annali dell'Istituto e Museo di Storia della Scienza, Firenze* 5, no. 2 (1980): 94; Leonida Rosino, "Il *Dialogo* come occasione per il processo e la condanna di Galileo," in *Giornate Lincee indette in occasione del 350° anniversario della pubblicazione del* Dialogo sopra i massimi sistemi *di Galileo Galilei, Roma, 6-7 maggio 1982* (Rome: Accademia Nazionale dei Lincei, 1983); articles in *Novità celesti e crisi del sapere. Atti del convegno internazionale di studi galileiani, 1982*, ed. Paolo Galluzzi (Florence: Giunti-Barbéra, 1984); Mario D'Addio, *Considerazioni sui processi a Galileo* (Rome: Herder, 1985); Mario Biagioli, *Galileo, Courtier* (Berkeley: University of California Press, 1993); Eugenio Garin, "Gli scandali della nuova 'filosofia'," *Nuncius* 8, no. 2 (1993): 417-31; and Paolo Simoncelli, "Galileo e la curia: un problema," *Belfagor* 48 (1993): 29-42.

4. Pietro Redondi, *Galileo Heretic* (Princeton: Princeton University Press, 1987, orig. Torino, Einaudi, 1983).

5. Biagioli, *Galileo, Courtier*.

6. See Massimo Bucciantini and Maurizio Torrini, eds., *La diffusione del copernicanesimo in Italia, 1543-1610* (Florence: Olschki, 1997).

7. Giorgio de Santillana, *The Crime of Galileo* (Chicago: University of Chicago Press, 1955); but see Massimo Bucciantini, *Contro Galileo: alle origini dell'"Affaire"* (Florence: Olschki, 1995).

8. Paul Lawrence Rose, *The Italian Renaissance of Mathematics: Studies on Humanists and Mathematicians from Petrarch to Galileo* (Geneva: Droz, 1975).

9. Charles B. Schmitt, "Filippo Fantoni: Galileo's Predecessor as Mathematical Lecturer at Pisa," *Science and History, Studies in Honor of Edward Rosen*, ed. E. Hilfstein et al. (Warsaw: Ossolinem, 1978), 53-62.

10. My information about Tartaglia is from Rose, *The Italian Renaissance of Mathematics*, 130; about Maurolico's manuscripts, Rosario Moscheo, "Scienza e cultura a

Messina fra Cinque e Seicento. Vicende e dispersione dei manoscritti autografi di Francesco Maurolico," *La rivolta di Messina, 1674-8 e il mondo mediterraneo nella seconda metà del Seicento, Convegno Storico Internazionale, Messina 10-12 ottobre, 1975,* ed. Saverio di Bella (Naples: L.P.E., 1978), 435-74.

11. Nicholas Copernicus, *De revolutionibus orbium coelestium libri sex* (Basel: Henricpetrina, 1566), from Copernicus' preface to Pope Paul III, n.p.: "Et quamvis sciam, hominis philosophi cogitationes esse remotas a iudicio vulgi, propterea quod illius studium sit, veritatem omnibus in rebus, quatenus id a Deo rationi humanae permissum est, inquirere: tamen alienas prorsus a rectitudine opiniones fugiendas censeo." Copernicus' elitist views about his audience are the subject of Robert S. Westman, "The Astronomer's Role in the Sixteenth Century: A Preliminary Study," *History of Science* 18 (1980): 105-47.

12. Niccolò Tartaglia, *Nova scientia inventa da N. T.* (Venice: 1537).

13. Leonard Fuchs, *Libri 4: difficulium aliquot quaestionum et hodie passim controversiarum explicationes continentes, magno studio aucti et recogniti* (Basel: Robert Winter, 1540); Leonardo Fioravanti, *Il tesoro della vita umana,* which I read in the edition of Venice: Spineda, 1629, instead of the original one of Venice: 1570.

14. The following remarks are intended to complement, not to contradict, the account by Paula Findlen in *Possessing Nature: Museums, Collecting and Scientific Culture in Early Modern Italy* (Berkeley: University of California Press, 1994), chap. 2.

15. Giovanni Cristofano Amaduzzi, ed., *Lettere italiane di alcuni illustri scrittori del secoli sedicesimo e diciassettesimo . . . cavate dai loro originali* (Rome: [1782]), 13, 18; Ulisse Aldrovandi to Card. Guglielmo Sirleto, 23 July 1577: "[Ho] sempre affaticato giorno e notte per illustrare questa praeclara filosofia naturale, avendo raccolte e ridotte insieme nel mio teatro di natura più di tredicimila cose diverse fra i quali si comprendono tutte le cose inanimate ed animate. . . . E di questo che io dico, se è vero o non, se ne può molto ben informar Sua Beatitudine; così dell'aggregazione delle cose da me raccolte in un luogo nello spazio di trent'anni, con tante gran spese ne' viaggi fatti." Postal routes are the subject of Bruno Caizzi, *Dalla posta dei re alla posta di tutti. Territorio e comunicazioni in Italia dal XVI secolo all'Unità* (Milan: Angeli, 1993); Eugène Vaillè, *Histoire des postes jusqu'à la Revolution* (Paris: Presses Universitaires Françaises, 1948), 31. In Venice, the Couriers' guild was established only in 1540, and its only major extraterritorial service was to Rome. Antonio Marzari, *Storia postale di Venezia* (Padua: Aulisio, 1976), 23. Aldrovandi's complaints about the mail are in Costanzo Felice, *Lettere a Ulisse Aldrovandi,* ed. Giorgio Nonni (Urbino: Quattro Venti, 1982), 43, 91.

16. *Lettere italiane,* 14: "Ponendo l'opinione mia in ciascuna particolar storia da me osservata, e non dagli antichi e moderni descritta, dichiarando a pieno l'istorie descritte dagli antichi, dando il giudizio di ciascuna cosa naturale secondo vari scrittori greci, latini, Arabi, ed in ciascuna di queste istorie, si corregono, ed esplicano molti luoghi di poeti, ed altri autori greci e latini, descrivendo ancora temperatura, e qualità di ciascuna cosa, dando parimenti notizia di tanti belli secreti di natura, i quali il grande Iddio per sua bontà infinita ha dato alla generazione umana a vari usi." 15: "Non c'è cosa sotto la luna che non si riduca a uno di questi tre generi, cioè alle cose inanimate, e fossili, che si cavano dalle viscere della Terra, o piante, o animali, ed ancorchè artificiali siano, inchiudino però in sua materia un di questi tre generi. Di modo che tutto il mondo sublunare di misti perfetti riluce in questi tre generi, de' quali noi particolarmente abbiamo scritte l'istorie."

17. *Lettere italiane,* 16: "Avrei bisogno di un gran principe mecenate, che mi desse una provisione onorata non per mio particolare, ma per benefizio universale, acciò che potessi aver molti scrittori, copiatori, scultori e pittori per dare in luce quanto prima queste mie onorate fatiche." And again, 18: "Tutto il giorno sono veduti da tanti vari signori, che passano per questa città, i quali visitano il mio pandechio di natura come un

ottavo miracolo del mondo."

18. Just one example of such resistance is in the comments of Giambattista Della Porta, *Magiae Naturalis libri XX* (Naples: Salviarno, 1588), preface n.p.: "Quae noxia et malefica obscuravimus, non ita tamen ut ingeniosissimus quisque detegere et percipere non possit, nec tamen clare, ut ignare turbae prostent, non tam occulte quin ingenium per quirentis accipiant, nec tamen aperte, ut in recessu eadem, quae in fronte promittant." I include him among the science practitioners because of what Gabriella Belloni says in "Conoscenza magica e ricerca scientifica in Giambattista Della Porta," in Idem, ed., Giambattista Della Porta, *Criptologia* (Rome: Centro Italiano di Studi Umanistici, 1982), 45-153. The nuances in the relation between science and humanism are explained by Cesare Vasoli, *Profezia e ragione* (Naples: Morano, 1974), 407-91, although not in relation to communication.

19. The standard study of Renaissance educational ideas is Paul Grendler, *Schooling in Renaissance Italy. Literacy and Learning, 1300-1600* (Baltimore: Johns Hopkins, 1989).

20. Daniele Barbaro, *I dieci libri dell'Architettura di Vitruvio* (Venezia: 1567), 445: "I nomi del nostro arsenale [sono usati] . . . acciò meglio si piglia la pratica di tai cose."

21. Alessandro Piccolomini, *Della filosofia naturale* (Venice: Giorgio dei Cavalli, 1565), 7: "È forza, se alcuna cosa vogliono sapere, di consumar in accenti, e vocaboli, e strutture loro, non solo la fanciulezza, e l'adolescenza, età attisime a far ferma impressione, ma la giovinezza ancora, e buona parte dell'età matura, che sono il nervo del nostro intelletto." Piccolomini's vernacular program is described in Alessandro Del Fante, providing ample bibliography, in "Amore, famiglia e matrimonio nell'*Istituzioni* di A. Piccolomini," *Nuova rivista storica*, 68 (1984): 511-24.

22. Piccolomini, *Della filosofia*, 10.

23. Sperone Speroni, *Dialogo delle lingue e dialogo della retorica*, ed. G. De Robertis (Lanciano: R. Carabba, 1912), 80-81.

24. Armando Serra, "Corrieri e postieri sull'itinerario Venezia-Roma nel cinquecento," *Le poste dei Tasso, un impresa in Europa*, contributi in occasione della mostra, I Tasso, L'Evoluzione delle Poste, Bergamo, 28 aprile-3 giugno 1984 (Bergamo: Comune, n.d.), 33-50; Vaillè, *Histoire des postes*, 46.

25. Consider Peter Dear, *Mersenne and the Learning of the Schools* (Ithaca: Cornell University Press, 1988); *1588-1988, quatrième centenaire de la naissance de Marin Mersenne: colloque scientifique international et celebration nationale Université de Maine* (Le Mans : Faculte des lettres, Universite de Maine, 1994); in addition, Anne Reinbold, ed., *Peiresc, ou la passion de connaître: colloque de Carpentras, novembre 1987* (Paris: J. Vrin, 1990).

26. Federico Cesi, *Del naturale desiderio di sapere et instituzione de' Lincei per adempimento di esso*, reproduced in Giuseppe Montalenti et al., *Federico Cesi e la fondazione dell'Accademia dei Lincei: Mostra Bibliografica e Documentaria* (Naples: Istituto Italiano per gli Studi Filosofici, 1988), 129, where he also recognizes the necessity to protect discoveries with the advance of the printing press. Cesi's proposal for a new type of correspondence is mentioned in *Edizione nazionale delle opere di Galileo Galilei*, 20 vols, ed. Antonio Favaro, 3rd edition (Florence: Giunti-Barbéra, 1967), vol. 11, 507, Federico Cesi to Galileo, 11 May 1613: "È parso necessario, in alcuni colloqui fatti questi giorni addietro, pensando all'accrescimento che è per seguire, di dare una norma allo scrivere delle lettere e loro titoli, poichè nasceranno occasioni spesse di scriver a molti e differenti e non praticati; e par che convenga alla purità filosofica, che deve professarsi, staccarsi affatto dall'usi aulici e ordinarii." The Lincei's correspondence is edited in G. Gabrieli, "Il carteggio Linceo della vecchia Accademia di Federico Cesi (1603-30), *Memorie del R. Accademia dei Lincei: Classe di scienze morali, storiche, e filologiche*, ser. 6, vol. 7, facs. 1, 2, 3 (1938-42), and analyzed, without due regard for the

problem of communication, by Giuseppe Olmi, "'In essercitio universale di contemplatione e pratica': Federico Cesi e i Lincei," in Ezio Raimondi and Laetitia Boehm, eds., *Università, accademie e società scientifiche in Italia e in Germania dal Cinque al Seicento* (Bologna: Il Mulino, 1981), 169-99; and Jean-Michel Gardair, "I Lincei: I soggetti, i luoghi, le attività," *Quaderni storici*, 16 (1981): 763-87.

27. Carl B. Boyer, *A History of Mathematics* (New York: John Wiley, 1968), 342-45; and especially, Charles Naux, *Histoire des logarithmes de Neper à Euler* 2 vols. (Paris: Blanchard, 1971).

28. Concerning the background to these problems, Owen Gingerich and Robert S, Westman, *The Wittich Connection. Conflict and Priority in Late Sixteenth-Century Cosmology* (Philadelphia: American Philosophical Society, 1988).

29. *Opere di Galileo Galilei*, 10: 365, letter dated 31 May 1610: "Oltre l'havere il Magini scritto al Matematico [Favaro's note: this is Zugmesser] di Colonia per tirarlo alla sua contro di lei, ha fatto il medesimo con tutti i matematici di Germania, Francia, Fiandra, Polonia, Inghilterra, ecc., il che ho saputo non da uno, ma da diversi di diverse nazioni."

30. The purely rhetorical aspects of this campaign are discussed in Maria Luisa Altieri Biagi, *Galileo Galilei e la terminologia tecnico-scientifica* (Florence: Olschki, 1965); Maurice Finocchiaro, *Galileo and the Art of Reasoning: Rhetorical Foundations of Scientific Method* (Dordrecht-Boston: D. Reidel, 1980); Jean Dietz Moss, *Novelties in the Heavens: Rhetoric and Science in the Copernican Controversy* (Chicago: University of Chicago Press, 1993).

31. *Opere di Galileo Galilei*, 5: 12.

32. *Opere di Galileo Galilei*, 4: 64, preface to the *Discourse on Floating Bodies*: "Lo scrivere [è] singolare mezzo per far conoscere il vero dal falso, le reali dall'apparenti ragioni, assai migliore che il disputare in voce, dove l'uno o l'altro, e ben spesso amendue che disputano, riscaldandosi di soverchio o di soverchio alzando la voce, o non si lasciano intendere o trasportati dall'ostinazione di non si cedere l'uno l'altro lontani dal primo proponimento, con la novità delle varie proposte confondendo lor medesimi e gli uditori insieme. . . . Perchè la dottrina che io seguito nel proposito che si tratta è diversa da quella d'Aristotele e dai suoi principi, ho considerato che contro l'autorità di quell'uomo grandissimo, la quale appresso di molti mette in sospetto di falso ciò che esce dalle scuole peripatetici, si possa molto meglio dir sua ragione con la penna che con la lingua." *Opere di Galileo Galilei*, 11: 334, to Giuliano de' Medici, 23 VI 12: "[In Italiano] acciò possa esser inteso, almeno in gran parte, da tutta la città, perchè così ha portato l'occasione di certa disputa, come nel principio dell'opera intenderà."

33. *Opere di Galileo Galilei*, 6: 219.

34. *Opere di Galileo Galilei*, 14: 49, to Elio Diodati, 29 October 1629. The specific rhetorical devices used by Galileo in this work, as well as its scientific content, are sufficiently analyzed by Maurice Finocchiaro, *Galileo and the Art of Reasoning*, correcting some perspectives of Stillman Drake, *Galileo at Work*; nonetheless he believes that the persuasiveness of a work can be judged entirely on the basis of the way it is put together without the slightest reference to what readers said about it at the time. I am also indebted to articles by Biagi and by Drake in *Giornate Lincee*.

35. From the Preface "Al discreto lettore," *Opere di Galileo Galilei*, 7: 32: "Non mancò chi temerariamente asserì, quel decreto esser stato parto non di giudizioso esame, ma di passione troppo poco informata, e si udirono querele, che consultori totalmente inesperti delle osservazioni astronomiche non dovevano con proibizione repentina tarpare l'ale a gli intellettuali speculativi. Non potè tacer il mio zelo in udir la temerità di si fatti lamenti. Giudicai, come pienamente instrutto di quella prudentissima determinazione, comparir pubblicamente nel teatro del mondo come testimonio di sincera verità. . . . Per tanto è mio consiglio nella presente fatica mostrare alle nazioni forestiere che di

questa materia si ne sa tanto in Italia e particolarmente in Roma quanto possa mai averne immaginato la diligenza oltramontana; e raccogliendo insieme tutte le speculazioni proprie intorno al sistema copernicano, far sapere che precedette la notizia di tutte alla censura romana." The best study of the Copernicus episode is Owen Gingerich, "The Censorship of Copernicus' *De Revolutionibus*," *Annali dell'Istituto e Museo di Storia della Scienza, Firenze* 6, no. 1 (1981): 45-61.

36. In the *Letter to Christina, Opere di Galileo Galilei*, 5: 310: "La gloria e la grandezza del sommo Dio mirabilmente si scorge in tutte le sue fatture, e divinamente si legge nell'aperto libro del cielo."

37. *Opere di Galileo Galilei*, 7: 129: "L'intelligenza divina ne sa bene infinite proposizioni di più, perchè le sa tutte, ma di quelle poche intese dall'intelletto umano credo che la cognizione agguagli la divina nella certezza obiettiva." Ecclesiastical censors in Rome later mentioned the quotation with horror, even though it was a commonplace already in sixteenth-century mathematical treatises.

38. *Opere di Galileo Galilei*, 11: 327, to Paolo Gualdo, 6 June 1612, in reference to his *Istorie e dimostrazioni intorno alle macchie solari* (Florence: 1613): "Io l'ho scritta volgare perche ho bisogno che ogni persona la possi leggere, e per questo medesimo rispetto ho scritto nel medesimo idioma questo ultimo mio trattato [*Discorso intorno alle cose che stanno in su l'acqua* (Florence: 1612)] e la ragione che mi muove, è il vedere, che mandandosi per gli Studi indifferentemente i giovani per farsi medici, filosofi, eccetera, sì come molti si applicano a tali professioni, essendovi inettissimi, così altri, che sariano atti, restano occupati o nelle cure familiari o in altre occupazioni aliene dalla litteratura; benchè come dice Ruzzante, forniti di un *bon snaturale*, tuttavia, non potendo vedere le cose scritte in *baos*, si vanno persuadendo che in quei *slibrazzon che suppie dè gran noelle de luorica e de filuorica, e conse purassè che strapasse in elto purassè.*" The italicized portions are Galileo's interpretation of the Paduan dialect of playwright Angelo Beolco, called Ruzzante, a personal friend.

39. *Opere di Galileo Galilei*, IV: 23, unpublished notes preparatory to the *Discourse on Floating Bodies*: "'E bene che il principe abbia filosofi discordi e di sette diverse, perchè così meglio si ritrova il vero; sì come per i medesimi è bene che i loro ministri siano discordi, ed i loro vassali in parte ed in inimicizie, perchè così hanno la robba, la vita e lo stato in maggior securtà."

40. *Opere di Galileo Galilei*, 4: 65, in connection with his *Discourse on Floating Bodies*: "Perciò che trattandosi se la figura de' solidi operi o non nell'andare essi o non andare a fondo nell'acqua, in occorrenze di fabbricar porti o altre macchine sopra l'acqua che avvengono per lo più in affari di molto rilievo, può essere di giovamento saperne la verità."

41. *Opere di Galileo Galilei*, (*Discourse on Floating Bodies*), 4: 177: "Troppo perderebbero di frequenza gli Studi e le scuole pubbliche, e poco sarebbero ascoltati i grandi insegnatori che hanno Aristotile per guida e per maestro . . . [se gli studenti] diviassero . . .
dalla strada piana e sicura delle filosofia peripatetica ad altra nuova, piena di rivolgimenti, e che sotto diverse facce rappresenta tutte le cose dell'Universo."

42. *Opere di Galileo Galilei*, 11: 327, from the same letter to Gualdo cited above: "Io voglio che vegghino che la natura, sì come gli ha dati gli occhi per veder l'opere sue, così bene come ai filuorichi, gli ha anco dato il cervello da poterle intendere e capire."

43. In the *Letter to Christina, Opere di Galileo Galilei*, 5: 310: "Il proibire tutta la scienza, che altro sarebbe che un riprovare cento luoghi delle Sacre Lettere, i quali insegnano come la gloria e la grandezza del sommo Dio mirabilmente si scorge in tutte le sue fatture, e divinamente si legge nell'aperto libro del cielo."

44. Sforza Pallavicino, *Trattato dello stile e del dialogo, ove nel cercarsi l'idea dello scrivere insegnativo, discorresi partitamente de' veri pregi dello stile sì latino come Italiano* (Rome: Mascardi, 1662 [1st impression: 1661]). All further citations will be

taken from this edition, which is an elaboration of his *Considerazioni sopra l'arte dello stile e del dialogo, con occasione di esaminare questo problema: se alle materie scientifiche convenga qualche eleganza ed ornamento di stile e quale* (Rome: Corbelletti, 1646), chap. 4

45. *Trattato dello stile*, chap. 30

46. *Opere di Galileo Galilei*, 19: 350, referral by Melchiorre Inchofer, 1632: "Munus docentis inter alias est, praecepta artis tradere, quae faciliora et magis expedita censet, ut faciles et dociles discipulos nanciscantur; proposita praesertim novitate disciplinae, quae curiosa ingenia mirifice solet allicere. In hoc genere, quam dextrum et solertum se praebeat Galilaeus, patet totum librum perlegenti." Concerning censorship, among others, Gigliola Fragnito, *La bibbia al rogo: la censura ecclesiastica e i volgarizzamenti della Scrittura: 1471-1605* (Bologna: Mulino, 1997); Mario Infelise, *Libri proibiti : da Gutenberg all' "Encyclopedie"* (Bari: Laterza, 1999).

47. *Opere di Galileo Galilei*, 19: 359, referral of Zacharia Pasqualigo: "Apporta la sua dottrina in tal maniera, che molti, anco intendenti nelle scienze matematiche, restano persuasi."

48. Paolo Sarpi, *Scritti giurisdizionalistici*, ed. Giovanni Gambarin (Bari: Laterza, 1958), 224-25

49. *Scritti giurisdizionalistici*, 223.

50. *Opere di Galileo Galilei*, 15: 56, Francesco Niccolini in Rome to Andrea Cioli in Florence, 27 February, 1633: "Sua Eminenza rispose [a Galileo], che le voleva bene e lo stimava per uomo singolare, ma che questa materia è assai delicata, potendosi introdurre qualche dogma fantastico nel mondo e particolarmente in Firenze, dove io sapevo che gli ingegni erano assai sottili e curiosi, massime che egli riferisce molto più validamente quel che fa per la parte del moto della terra che quel che si può addurre per l'altro. "

51. Pallavicino, *Trattato dello stile*, chap. 35: "Talora ultimamente il fine loro è di mostrare la debolezza delle prove comuni, e l'oscurità de' problemi che il volgo animosamente risolve per evidenti; acciò che s'accenda nei lettori l'avidità di speculare con sottigliezza."

52. *Opere di Galileo Galilei*, 7: 143.

53. *Opere di Galileo Galilei*, 14: 366.

54. *Opere di Galileo Galilei*, 11: 99, 6 May 1611: "Che la terra giri, sinora non ho trovato nè filosofo nè astrologo che si voglia sottoscrivere all'opera di Vostra Signoria, e molto meno vorranno fare i Teologi. . . . A me pare, che gloria s'abbia acquistata con l'osservazione della luna, nei quattro pianeti, e cose simili, senza pigliar a diffendere cosa tanto contrario all'intelligenza e capacità degli uomini, essendo pochissimi quelli che sappiano che cosa voglia dire l'osservanza dei segni e aspetti celesti."

55. Piccolomini, *Della filosofia*, 9: "Non vorrei . . . che i letterati si pensassero d'haver a legger [i miei libri] come si leggono historie, o novelle; talmente che aprendo il libro dovunque si abbatti in leggendo. . . . [Piuttosto], cominciando dal principio, e seguendo di mano in mano, secondo che le cose tra di loro incatenate l'una dall'altra dipendono."

56. *Opere di Galileo Galilei*, 7: 482: "Farò ogni sforzo per lasciarmi intendere, ma la difficoltà dell'accidente stesso, e la grande astrazione di mente che ci vuole per capirlo, mi sgomentano."

57. From his *Risposta alle opposizioni di Lodovico delle Colombe e di Vincenzo di Grazia contro il trattato delle cose che stanno in su l'acqua* (Florence: Giunti, 1615), reprinted in *Opere di Galileo Galilei*, 4: 462: "[Il] vulgo . . . per non intendere i sensi delle scritture, si quieta sul vedere i caratteri e sul poter dir che sia stato risposto." Compare, in the same work, 4: 460. Similar ideas about problems of communication to the general public on other occasions occur in 5: 310, 462.

58. *Correspondence de Marin Mersenne*, vol. 6, ed. Paul Tannery and Cornélius De Waard (Paris: Éditions du CNRS, 1960), 201, Christophe Villiers in Sens to Mersenne,

24 February 1637: "Il doit bien y avoir des difficultez à soudre, esquelles Aristote, n'y presque toute l'Antiquité n'a posé, se contentant de suivre un grand maistre comme un torrent ou quelqu'autre premier mobile."

59. Indifferent was memorialist Giacinto Gigli, *Diario Romano, 1608-70* ed. Gino Ricciotti (Rome: Tuminelli, 1958). In addition, here, *Opere di Galileo Galilei*, 14: 103, from Antonio Badelli's "Avvisi" for 18 May 1630: "Qui si trova il Galilei, che è famoso matematico e astrologo, che tenta di stampare un libro nel quale impugna molte opinioni che sono sostenute dai Gesuiti. Egli si è lasciato intendere che dal D. Anna [Colonna, wife of Taddeo Barberini, of the Papal family] partorirà un figlio maschio, che alla fine di giugno havremo la pace in Italia e che poco dopo morirà D. Taddeo e il Papa. L'ultimo punto vien comprovato dal [Giovambattista] Caracioli Napoletano, dal P, [Tommaso] Campanella, e da molti discorsi in scritto, che trattano dell'elezione del nuovo Papa come se fosse sede vacante."

60. *Lettere inedite*, 2: 249, 5 September 1674: "Non è nemmeno infarinato negli elementi geometrici; non ha alcuna notizia delle materie che oggi si trattano per le accademie, non ha veduto alcun libro moderno, e sa il titolo; ma non ostante discorre del Galileo e della di lui abiura, e si dichiara inclinato al sistema copernicana, non per altro se non perchè, dice egli, non può capire quella gran velocità del primo mobile Tolemaico."

61. Alessandro Tassoni, *Lettere*, 2 vols., ed. Pietro Puliatti (Bari: Laterza, 1978), 2: 259. In general, Luigi Pepe, *Copernico e la questione copernicana in Italia* (Florence: Olschki, 1996).

Chapter 2:
The Business of Astrology

Just when Galileo's campaign for the heliocentric theory was reaching its climax, another stargazing natural philosopher, much less known, aroused the implacable ire of the Roman officials. Abbot of Santa Prassede and one-time general of the Vallombrosa Order, a branch of the Benedictines, Orazio Morandi was accused in 1630 of magic, fortune telling, and political chicanery. His most serious crime was to have predicted the death of Pope Urban VIII, based on certain astrological portents, and to have allowed news about this prediction to spread as far as Spain. Thereupon the Spanish cardinals embarked for Italy in order to attend a conclave that was not in fact to take place for another fourteen years—an embarrassment to themselves as well as to the pope. Urban VIII, furious at what he regarded as an act of political as well as astrological effrontery, personally ordered a criminal inquiry. The consequences were enormous. Astrology itself was hurt by the growing impression that predictions were mere merchandise in the game of political and social favors. Soon after the trial, Urban VIII promulgated the most severe anti-astrology legislation yet. And this legislation, as well as Morandi's crimes, may well have affected the outcome of the Galileo case, tried soon afterwards.[1]

For Morandi as for Galileo, the high-stakes patronage environment of the early seventeenth century demanded new market strategies. And for both, the experiment with them ended in disaster. Nor should we be too surprised that the market strategies of an astrologer occasionally coincided with those of an astronomer.[2] So far we have tried to represent the early seventeenth-century cultural world according to the terms set by Galileo: as a battle between him and his adversaries. By portraying himself as the enemy of Aristotelianism, Scholasticism, and any form of dogmatism, he also distanced himself from forms of dogmatic thought that have come down to us with names like "astrology" and "alchemy."

To contemporaries of Galileo the differences by no means seemed quite so clear. He himself appeared to accept at least one dogmatic preconception in his work, namely, the proposition that mathematical laws were prior to experience.[3] And on more than one occasion he was known to dabble in astrology.[4] Moreover, just as he enhanced the value of his contributions by producing them according to the criteria of a methodological revolution he was carrying on in the fields of natural philosophy that interested him, likewise, astrologers and other practitioners in the collateral fields of natural knowledge attempted to enhance the value of their contributions by submitting them as products of a similar methodological revolution in their respective fields. While Galileo appealed to a wider intellectual marketplace to gain support for his methodological innovations, astrologers tried to do the same.[5] In this chapter we will examine the role of sheer intellectual enterprise in defining the characteristics of seventeenth-century natural knowledge. In the next chapter we will see how it benefited Galileo's disciples.

No market, of course, was entirely free in the early modern period.[6] Commodities signified values in terms of honor and prestige as much as values in terms of money. Since the social status of the acquirer of a commodity affected its value as much as did the status of the producer, commodities were inscribed as much within circuits of personal favor as within circuits of exchange. Practitioners were well aware that adherence to specific codes of behavior was essential to their success. This is the sense in which we may speak of a patronage marketplace in this period. And careful attention to the exigencies of the marketplace marked the astute cultural entrepreneur.

Orazio Morandi focused his entrepreneurial zeal on attaining a position of greater eminence than his origins among the minor ranks of the Roman patriciate might otherwise permit. Once he had risen through the hierarchy of the Vallombrosa order, in Florence and then in Rome, he considered his carefully cultivated Medici ties to have yielded their best fruit. He thereupon set out to perfect the intellectual and cultural techniques necessary for attracting greater attention to himself. His choice, no doubt informed by the examples of such successful sixteenth-century astrologers as Luca Gaurico and Girolamo Cardano, fell upon the occult sciences.

To build an image of professional competence sufficient to support his claims to mastery of the occult, Morandi expanded the existing library of his monastery in Rome. It was already well equipped with a repertory of Italian incunabula as well as a rich sampling of sixteenth-century titles in the humanities, classics, and theology from all over the peninsula and the rest of Europe as well—not to mention key humanist manuscripts going back to the founding of the monastery in the thirteenth century.[7] Morandi put his own stamp on the library between 1614 and 1630, when he was abbot of the monastery and, at various times, general of the order and its solicitor at the papal court.

To draw attention to his book acquisition as well as to widen the circle of those who might be interested in his astrological expertise, Morandi conceived of a unique expedient: he opened the library up to the Roman community—or, at least, to parts of it—as a public library. What is more, he allowed certain borrowers to take books home. And in fact, one of the most remarkable documents in the trial record is the complete library lending list for three years in the 1620s, one of the earliest such documents we know, attesting to the circulation of books of scholarship and literature to an audience that included priests and laymen, artists and patrons, writers and scribes, students, and women.[8]

By the time Morandi got through with it, the library contained a particularly rich collection in matters of all sorts relating to the heavens. And typically, in the early seventeenth century, matters relating to the heavens largely included matters relating to astrology. So the library not only possessed editions (and we are often able to tell which ones) of standard astronomical classics, ancient and modern, from Aristarchus, Hipparchus, and Ptolemy to Copernicus, Galileo and Kepler, as well as the main sets of planetary tables and ephemerides.[9] It also possessed a complete set of astrological works, from Ptolemy, Julius Firmicus Maternus, and Aratus to the medieval Arabic author Abenragel, to more recent works by Henricus Lindhout, Henrik Rantzau, and Rudolph Goclenius.[10]

Astrology, of course, was no less dangerous in the seventeenth century than was astronomy.[11] We do not know whether the Santa Prassede library copy of Copernicus' *On the Revolutions of the Heavenly Spheres* had ever been "corrected" as the 1616 papal decree demanded. What we do know is that works by astrologers and astrology sympathizers Joachim Camerarius, Philip Melanchthon, and Cyprian Leowitz, notoriously ill regarded by the Index, were present in the library and customarily lent out.[12]

In spite of all the efforts of astrology's apologists to defend such interests against ecclesiastical interference, Sixtus V's pronouncement was still as valid in the seventeenth century as when he had first made it in 1585.[13] To wit, condemning all "astrologers . . . who claim to have a vain knowledge of the stars and planets and most audaciously purport to foresee the divine dispositions in their time." Such persons, it explained, "draw up nativities or genitures concerning the motion of the planets and the courses of the stars, and judge future and even present affairs, especially occult ones." They presume, from the birth of the child and its natal day, "and any other most various observation and notation of any time and moment," to foretell "their condition, course of life, honors, wealth, children, health, death, journeys, struggles, enemies, imprisonments, assassinations, and various dangers and other adverse or prosperous occurrences or events." And it threatened severe action against anyone "who knowingly reads or keeps these sorts of books and writings or such that contain these matters."

Thus, when Abbot Gherardo Gherardi, visiting Rome from the University of Padua where he taught philosophy, borrowed a "manuscript of nativities of cardinals" from the Santa Prassede library on 13 March 1629, as our library log informs us, there was no doubt that such works came under the ban, no matter which cardinals' nativities were being described or for what purpose. Other astrological manuscripts that circulated could be dangerous merely by virtue of their authors' reputations. One very dangerous reputation was that of Tommaso Campanella, whose Inquisition trial was still going on. Around that time Francesco Usimbardi, a prelate in the Apostolic Chamber, borrowed an unspecified "manuscript of Campanella" from Santa Prassede, perhaps the same manuscript "astrology" that was borrowed by another library patron named Stefano Senarega about whom we have no other information.[14]

To explain the appeal of judicial astrology in a city where some of the great discoveries of modern astronomy were being discussed, there is probably no need to repeat what Keith Thomas has already written concerning the troubled times and the difficulties of life.[15] Other inducements no doubt included the uncertain fortunes of courtiers and clients of wealthy patrons. Roman society was special in this respect, because rewards and prizes were distributed at the whim of an absolute ruler, usually elected in his declining years, upon whose death an entirely new regime was customarily erected. Even in the upper echelons of Roman society, fortunes were made and broken every day.[16]

To this explanation for the perceived increase in astrological activity after the turn of the seventeenth century we might even add the fin de siècle syndrome. That the year 1600 had passed without a universal cataclysm put no more of a damper on such speculations than had the failure of the year 1500 to yield the same, or, for that matter, the year 1524, when the great conjunction of the three upper planets was supposed to occur in the sign of Pisces.[17] As late as 1603, Tommaso Campanella was still as convinced as Cyprian Leowitz had been in 1564 that the year 1600 signaled the beginning of a new age. The spectacular nova that followed the great conjunction of 1603 only confirmed him in this belief.[18] In 1628, the Sicilian virtuoso Giambattista Hodierna named six ages since Creation, beginning with the age of gold and proceeding through the elements from silver to copper to iron before finally arriving at the age of lead in which firearms held sway. Just begun, he asserted, was the age of crystal, ushered in when the telescope drew us to the sphere beyond the planets, a sphere greater in nobility, prestige, and sublimity than all the others. In this age, he proclaimed, human intellect would triumph over bestiality and grossness. "Just look around at the admirable acuteness of the human mind that we observe in everyone," he pointed out, perhaps with some exaggeration, "whereas each person is able, in an unutterable way, to penetrate into the intimate recesses of the mind of another."[19]

How much this new demand for astronomy contributed to the new vogue astrology began to enjoy in this period is difficult to assess. In Rome, at least, an astrological resurgence was widely thought to be under way. And the rather uncharitable witness in Morandi's trial who stated to the court "all Rome is filled with these charlatans" was by no means alone.[20] Surely the "various Spaniards," a "knife seller" and "a certain Battelli," whom he named as examples, would have fully agreed concerning the prevalence of the practice, if not the negative characterization. Nor was another witness exaggerating when he noted, "astrology has become a recognized profession, and almost everyone has nativities drawn up. . . . [Indeed,] there is no cardinal or prelate or prince who does not have discourses drawn up telling his fortune based on his nativity. . . . Of course," he was quick to add, "only for secondary causes, because everything in the final analysis depends on God."[21]

If astrologers could be punished at any time, and were, why did the Roman authorities leave the library and its custodians alone for years before pouncing on poor Morandi? One reason is some of the very people who might have ordered action to be taken against the monks were themselves some of the most dedicated members of the library circle. For instance, Cardinals Tiberio Muti, Luigi Caetani, and Luigi Capponi, besides their exalted prelatic status, also happened to be sitting on one of the most important committees in ecclesiastical government—namely, the Congregation on the Index.[22] And just in case their authority was not enough to legitimize the monastery's activity, there was also Nicolo Ridolfi, the Master of the Sacred Palace, whose signature was required on every permit to publish a book.[23] Moreover, when Cardinal Scaglia and Francesco Usimbardi, as well as Francesco Maria Ghisleri, an auditor of the Rota, all came looking for astrological advice at the monastery it was freely given, and their nativities, along with those of many other library patrons, were duly recorded in the monks' astrological notebook, which has survived in the trial records.[24]

To persuade clients such as these about the value of his expertise, Morandi determined to refound the discipline according to more modern scientific criteria. He was, of course, not the first to try this. Ever since Pico della Mirandola expressed his famous objections to astrology in the late fifteenth century, adepts had fought back. With the advent of a more empirical and experience-oriented approach to natural philosophy in the latter half of the sixteenth century, the effort to defend astrology became more empirical and experience-oriented—not to say, more voluminous. Consider Luca Gaurico and Girolamo Cardano, who both tried to prove the validity of their different forms of astrology by a selection of some one hundred nativities apiece, with the pertinent explanations for how they illustrated the lives in question, and compare them to Johannes Garcaeus, in the last third of the century, who included no less than four hundred nativities.[25] Meanwhile, Rudolf Goclenius tried to apply to divination the same empirical methods that the astrologers were applying to astrology, by observing carefully, for instance, that a sampling of men on whose foreheads he had noted a peculiar cruciform mark included fifty-one who died violent deaths.[26]

For Morandi, scientific astrology meant very much the same as scientific astronomy: a body of knowledge based on proven facts. And, at times, like most of his contemporaries, he found the two disciplines very hard to tell apart. Not only did well-regarded ancients like Ptolemy write on both subjects; even the most serious modern compilers of astronomical information, such as Erasmus Reinhold, author of the Prutenic Tables of planetary motions, larded their works with astrological predictions of various types. No wonder major medical faculties were just as likely to have chairs devoted to one as to the other. And no wonder Morandi situated astrological and astronomical texts side by side in his library. Like most of his contemporaries, he would have heartily agreed with Kepler's characterization of astrology as astronomy's "mother and nourisher," unable to "deny her beloved daughter" anything.[27]

Whatever the family relation was between astrology and astronomy, developments within the latter discipline inevitably rubbed off on the former. And in many ways, as we saw in the last chapter, this was astronomy's finest hour. As the new astronomy developed, the best practitioners of astrology hurried to apply the latest notions to their discipline. Ilario Altobelli compiled a new and more accurate (so he claimed) set of tables of planetary movements. On the basis of these, he attempted to work out a more exact division of the twelve houses than had hitherto been possible because of the irregularities produced by the obliquity of the ecliptic with respect to the celestial equator, causing the different houses to have different lengths. In rejecting the equal house method and eight other methods currently in use, Altobelli claimed to "confound the followers of Regiomontanus"—not to mention those of Cardano.[28] And had his ideas not been stolen by Andrea Argoli and incorporated into the latter's latest *Ephemerides*, so he complained to Morandi, he would surely have won credit for having restored the reputation of Ptolemy, who, in his view, came closest to the truth.[29]

And it was Francis Bacon, the influential theorist of the new approach to nature, who set out to clean the Augean stables of astrology once and for all. In passages of his *Advancement of Learning* that can now be safely taken out of the closet that a Whiggish historiography once put them in, he called for undertaking a new compilation and sorting of data: "The astrologers may, if they please, draw from real history all greater accidents, as inundations, plagues, wars, seditions, deaths of kings, etc., as also the motions of the celestial bodies . . . to . . . erect a probable rule of prediction." Such information was of course to be carefully scrutinized. "All traditions should be well-sifted, and those thrown out that manifestly clash with physical reasons, leaving such in their full force as comport well therewith." He never questioned the planetary influences themselves, upon which astrology was based—i.e., "the universal appetites and passions of matter" that constituted "those physical reasons [that] are best suited to our inquiry," along with "the simple genuine motions of the heavenly bodies." This, he avowed, "we take for the surest guide to astrology."[30] The monks could not have put it better themselves.

Greater rigor implied no claim to infallibility. On the contrary, Morandi readily admitted that there was "great uncertainty in this profession," beginning with "the principle foundations concerning the movements of the planets," about which "there are still many uncertainties."[31] After all, the "Medici planets" circling around Jupiter now joined new more distant bodies like Nova Cygni and Nova Ophiuchi to form a celestial canopy far different from any imagined by the Ancients, and likely to change still more. Another level of uncertainty came from coordinating a nativity with the planetary motions. Ptolemy suggested using the moment of the subject's conception in the womb as the basis—not an easy thing to discover, then or now. Other authors suggested various other procedures, but, said Morandi, "the effects do not correspond to their teachings." The challenge of a scientific astrology was to reduce the uncertainties by accumulating data, running tests and finding out which theory best saved the appearances.

Interrupted in the planning stage by the court's investigations, Morandi's efforts would have produced a comprehensive astrological encyclopedia of truly Baconian proportions for in-house consultation. As it is, we can only try to reconstruct, from notes scattered throughout the trial record, the way he hoped, for instance, to bring together the technical expertise not only of the monks at Santa Prassede but also of their neighbors. Thus, Ottavio Marini would supply information about purely astrological matters, such as the Part of Fortune, horizon lines and arcs, latitudes across the equator, solstice points, the aspects, the theory of the houses according to Ptolemy and according to Regiomontanus, and the theory of directions according to both.[32] Another neighbor, one P. Giambattista di Giuliano, would supply more properly astronomical expertise on questions concerning meridians, the theory of the planets, the equation of natural movements, the computations of the ecliptic, the measurement of the stars, the motion of the sun at the equator, triangulation, parallax, whether right ascensions increase with their greater declination from the ecliptic, and so forth. He too was responsible for answering some hard questions concerning the recent work of Tycho Brahe, such as why the meridian at Padua, according to Tycho's computations, was a full four minutes different from that of nearby Venice.

Despite the scientific approach, Morandi had no intention of abandoning his humanist background. Scholarly methods had not yet proceeded so far beyond the Renaissance for a fundamentally philological and literary approach to be excluded from a scientific endeavor of the sort he had in mind. Anyway, investigations and observations required some principles of organization or orientation if they were not to result in an undifferentiated morass of miscellaneous information. By referring to an authoritative and revered ancient author, he could find such principles without any more fear of compromising his own freedom to innovate than had Giacomo Zabarella and others among the self-proclaimed Aristotelians of their time. The best of the latter sort of philosophers, after all, referred to the Aristotelian tradition as a template and a background for original work.[33] So Morandi planned to utilize Ptolemy's *Tetrabiblos*, conveniently translated into Latin by Egidio Tebaldi and later by Joachim Camerarius, Philip Melanchthon, and Antonius Gogava, in the same way that Zabarella used the works of Aristotle.[34]

The choice of Ptolemy was hardly surprising. He still offered the best manual of basic astrology in any language. And the continuing effectiveness of Ptolemy's astrological system as a whole had no more to do with the condemnation of Copernicus than it did with the propagandistic efforts of thoroughgoing Ptolemaic astronomers like Christopher Clavius.[35] For all Clavius was able to do was to keep the master's ideas in circulation among colleagues at the Jesuit Collegio Romano in spite of the decided preference there for Aristotle. More importantly, even as Ptolemy's astronomy, along with that of Aristotle, and their geocentric views of the universe, began to lose ground inexorably to the heliocentric account, most recently proposed by Copernicus and Galileo, Ptolemy's view of the mechanisms for the influence of the planets on the terrestrial world remained unsurpassed.[36]

Indeed, over the years, Ptolemy's view showed extraordinary adaptability to late antique and medieval physiological conceptions that were still alive and well in the seventeenth century. He claimed that the planets exerted their influence by way of the four chief qualities of hot, cold, moist, and dry, which were central not only to the Hellenic cosmos he had inherited but also to the version of Galenic medicine that continued to hold sway in much of early modern Europe. The sun, he believed and the physicians confirmed, exerted a heating and drying action, while the moon acted by cooling and moistening, and so on in different combinations for each planet. The angles and modifications of these influences determined the kind of effect the planet might have—and the challenge to the astrologer was to figure this out. By the sixteenth century, Ptolemy's view was combined with other late antique ideas concerning the four humors, yielding the four major personality types—melancholy, sanguine, bilious, phlegmatic—that had been further developed in the Middle Ages. And the theory of Galen and Hippocrates on critical days and on the patterns of ebb and flow of diseases, which enjoyed an extraordinary vogue long before Agostino Nifo's widely read works on the subject, was viewed in a distinctly Ptolemaic light, if only because of Ptolemy's association with it through the spurious *Centiloqium* that circulated under his name.[37]

Few categorically denied some sort of planetary influences, and even those who referred to them only vaguely most probably had the ones described by Ptolemy in mind. Disagreement with Ptolemy's notions about the organization of the cosmos did not prevent Galileo from suggesting that the planets "abounded in influences"—though he never specified of what sort.[38] Even the opponents of Ptolemy's astrological ideas, such as Jofrancus Offusius, who tried to offer his own system, nevertheless accepted Ptolemy's account of planetary influences without too many modifications. Robert Fludd, on the radical fringe of natural philosophical reform, reinterpreted the Book of Genesis along alchemical lines and reformulated the story of Creation in order to end up with planets of the sort and with the same sorts of qualities and influences as those described by Ptolemy.[39]

Kepler built on Ptolemy's ideas in extraordinarily creative ways. At first he claimed that the beneficent influence of Jupiter, Venus, and Mercury was due to the structurally stable triangular faces of the polygons that, according to his calculations, yield the planetary distances from the sun:

> Mars, however, agrees with Saturn in evil alone. I therefore take this into consideration with the instability of their angles, which I find common to both. An opposite argument would thus suggest good qualities, that is, considering the stability of the angles at the bases. This is why Jupiter, Venus and Mercury are beneficial.[40]

Later he claimed that that the planetary aspects or angles between planets in the zodiac (i.e., opposition, trine, etc.) were not effective in themselves as in Ptolemy's theory, but only in so far as they affected the world soul, which in turn influenced the operation of all things in the universe. In eclipses, this world soul was "strongly disturbed by the loss of light."[41] The configurations of the stars at birth reminded individual souls of their "celestial character" and endowed them with the particular features that accompanied the subject through life.[42]

Morandi hewed very closely to the Ptolemaic line even in his discussion of the categories of the objects in the heavens. Just as in Ptolemy's Book 1, chaps.

1-16 (in Gogava's version), he organized his treatise so that the division of the planets into masculine and feminine, diurnal and nocturnal, would be followed by the division of the year into seasons and of the horizon into four angles.[43] Next would come the division of the signs of the zodiac into solstitial and equinoctial, solid and bicorporal, masculine and feminine, commanding and obeying. He would concur with Ptolemy's critique of the Egyptian concept of terms, or termini, i.e., places mapped out within each sign assigned to each of the five planets (excluding the sun and moon), from which the length of life could be calculated. He would also share Ptolemy's skepticism about the Chaldean method of calculating terms, which differed from the Egyptian one only in assigning greater predominance to the maleficient planets Mars and Saturn. In the end, rather than suggesting Ptolemy's own alternative to the Egyptian and Chaldean methods, he would follow Cardano in rejecting the idea entirely. But he would follow Ptolemy in taking account of the so-called applications, or planets that preceded other planets (in other words, lying to the west of them), said to "apply" to the latter, not to be confused with planetary aspects, which concerned not the real presence of the planet but its virtual presence by way of an occult connection.[44]

Morandi's discussion of the basic features of an astrological chart promised to stray a little further from the model. Following most contemporary astrological manuals, he would collect into single chapters the information that Ptolemy scattered here and there, and he would add to it whatever else he had in his files. First of all, he would distinguish the major beneficient planets, i.e., Jupiter, Venus, and the moon, so-called because of their moderate qualities of heat and moisture, from the major maleficients, Saturn and Mars, with their excessive cold and dryness; and all of these they would distinguish from the earth and Mercury, which could go either way. Then, one by one, he would work through the list of heavenly bodies, beginning with the sun, the lord of action, according to Ptolemy's system—i.e., the indicator in whose presence other planets determined the particular profession or way of life of the subject. And concerning the moon's nodes, he did not follow Ptolemy in rejecting them as pseudo-planets and useless accretions; instead, he embraced the view of the ninth-century Baghdad scholar Albumasar, who swore by their effects.[45] Likewise, in the discussion of Ptolemy's list of five chief aspects or angular relations between the signs or planets, namely, the conjunction, opposition, trine, quartile, and sextile, he would follow Kepler in adding four new aspects—to wit, the semiquadrate, quintile, sesquiquadrate, and biquintile, referring, respectively, to angles of 45, 72, 135, and 144 degrees. On comets and eclipses, he followed the Zeitgeist in placing far more emphasis on them than did Ptolemy, who dismissed them in a couple of sentences. Yet Morandi was no more certain than many contemporaries about what to do with the many new fixed stars that had appeared in the sky since Ptolemy's time, visible by the naked eye and by telescope. So he proposed only to discuss the influence of the ones mentioned by Ptolemy in the Tetrabiblos—Aldebaran in the constellation Taurus, Beta in Leo, Spica in Virgo, and so forth.

In the actual science of prediction, Morandi's adherence to Ptolemy may have been more apparent than real. After a discussion of the Part of Fortune he would add more material corresponding to the contents of Ptolemy's Book 3, chapters 1-12 and 15-18 in Gogava's translation, concerning the length of life, the form and temperament of the body, and the quality and the illnesses of the soul. And he would cover the standard Ptolemaic questions concerning the prediction of wealth, heredity, dignity, marriage, children, friends, and the ages of man, roughly corresponding to Book 4 of the *Tetrabiblos*. But in an age and environment far more preoccupied with individual fortunes than with collective ones, and more sensitive to adults' misfortunes than to children's, he left out entirely the branch of astrology allotted by Ptolemy (in his Book 2) to the interpretation of the fortunes of countries and peoples. And where Ptolemy followed this with a discussion of individual fortunes, beginning with children not reared, i.e., due to death at birth or abandonment by the mother, Morandi would instead discuss the death of the mother.

Morandi did his most original work when trying to discover recurrent characteristics in large numbers of natal charts. To test the various theories connecting stars to outcomes using scientific method he began compiling what might have become one of the most elaborate bodies of empirical evidence to date, had he not been sidetracked by the trial. In order to acquire the essential information he naturally relied on research—not only from printed sources, but from many other kinds of sources as well. To organize the collaboration in true Baconian fashion, he ordered his collaborators to draw up questionnaires with particular regard for ascertaining exact birth times to the nearest minute, always a controversial item in the debate about astrology's effectiveness.

Even collecting birth times had its hazards, Morandi noted. Baptismal records and "books of the dead," which could be found in every city, sometimes only cited an individual's year of birth; and the most accurate ones only went so far as to add the day. If all else failed, he decided to ask the parents, if living—although memories could be weak. "It should be noted that many times ordinary people may recall the month and the day and the hour, but most of them do not know the exact time, saying only that 'such and such was of such and such an age when he died,' often missing a year."[46] He declared only one source to be absolutely out of bounds. On no account were the collaborators to ask any astrologers—because of the latter's well-known proclivity to "accommodate things to their own purposes." No one knew this better than Morandi himself.

Afterwards, Morandi would sort the nativities according to outcome. And although he planned a section on individuals who obtained good fortune by evading dangers, as well as one on the varieties of fortune in general, what he presented for the most part amounted to a grim catalogue of possible catastrophes. In the first section he would place the cases of persons who fell from greatness or whose prosperity and happiness were interrupted by ill fortune including exile, pursuit by enemies, and wounds, as well as those who failed to gain an inheritance or other expected riches. Then he would give cases regarding things lost due to various disturbances, persons forced into mendicancy, and persons placed before the tribunal. This he would follow by a subsection on illness, including apoplexy, podagra, calculi, fluxes, spitting blood, and ruptured veins. Next came a subsection on death by the hand of justice, including by strangulation, by iron, by contusion, etc., and a subsection on death by another's hand including by iron, by fire, by poison, and by other means. In a final subsection on accidents, bearing witness to the triumph of fortune and the defeat of humankind, they noted death by fire, poison, falling from heights, ruin, and drowning.

If these investigations were intended to gain credit for astrology in a period of scientific change, while distinguished visitors shielded the monastery's activities from prosecution according to the various papal anti-astrology bulls, the strategy appears to have worked brilliantly. Morandi was able to invite the greatest cardinals in Rome over for his monastery soirées. The company occasionally included Galileo, a Morandi friend from schoolboy days in Florence, who was now busy seeking support in Rome for the publication of his *Dialogue Concerning the Two Chief World Systems*. The monks themselves provided entertainment, including music and theatrical presentations.[47]

But scientific astrology was not enough for achieving the glory and protection Morandi sought. He needed an outstanding application of his techniques. There were plenty of examples, in the fast-paced intellectual environment of the time. Natural philosophers like Galileo Galilei competed for the attention of powerful patrons by spectacular coups de scène.[48] Clearly, preeminence in astrology belonged to the successful bearer of an important prediction—a prediction about the future of an individual whose position in the firmament was of great interest not only to the usual satellites, but also to others of truly planetary importance. Finally the chance came, so Morandi thought, in 1630, and the object was none other than Urban VIII.

Predicting the death of a personage as eminent as Urban VIII, of course, was fraught with difficulties. For one thing, the prediction might fail. Only three years before, the pope's health seemed to be so sound that a previous Venetian ambassador saw no reason not to expect a "long pontificate."[49] However improbable a prediction of death might be, it was almost certain to encounter strong disapproval from the personage in question. Julius Firmicus Maternus was by no means the first astrologer in history to refuse to draw up nativities for the eminent because of their capacity for retaliation. Nor was Luca Gaurico the last to suffer for having done them. Half a century after Gaurico went to the torture chamber for having predicted Guido Bentivoglio's problems with Pope Julius II, the English astrologer Thomas Harriot was thrown in jail for having cast nativities for the king and crown prince on the eve of the Gunpowder Plot that narrowly missed destroying them both.[50]

Of all possible subjects for astrological prophecies, Urban VIII was perhaps among the most sensitive. Even those who knew him well admitted that the rumors about his singular dedication to this science were partly true. Two years before, when some of the Roman astrologers, including Francesco Lamponi, a member of the Santa Prassede circle, had foreseen an imminent threat due to a solar eclipse followed by a lunar eclipse, the Tuscan ambassador reported, "the pope has begun to calculate his nativity more than ever."[51] To counter the fatal astral influences, he called in none other than astrology expert Tommaso Campanella, who happened to be enjoying a rare period of relative freedom from the prisons of the Inquisition. With Campanella's help he had a room of Castel Gandolfo fitted out as a veritable magic chamber, decorated with dried reeds and white linen sheets, perfumed with burnt essences of laurel, rosemary, and cypress, and illuminated with burning torches in circles on the floor representing the spheres of the universe. Only Campanella's own later testimonies stand as proof that what went on in that room contained nothing to arouse suspicions of witchcraft or necromancy.[52]

Still, death was one area where the astrological teachings studied at Santa Prassede left little room for doubt—if the necessary data could be obtained. Of course, interpretation of the data was no easy matter; and Ptolemy, upon whom Morandi and his fellow-monks often relied, had added endless complications to the earlier Greek idea that the birth planet determines lifespan.[53] Instead, he calculated the lifespan based on the geometrical distance between two highly variable and movable points in the zodiac, beginning with the aphetic, also called the prorogating or life-giving point, where the planetary giver of life was situated. This could be in a number of different places, most properly in the Midheaven, but also in the ninth house. The planetary giver of life itself was usually either the sun or the moon, but not necessarily. The natural life span according to this theory was calculated from where the planetary giver of life was situated in the life-giving point, as far as the descendant (i.e., the opposite side of the nativity from the ascendant), unless interrupted by an anaretic planet, i.e., a planet with death-giving properties. Such planets might include the quintessentially unfortunate planets Saturn or Mars; they might include the moon if the sun happened to be the life-giving planet. A sign could also be anaretic—especially when in a quadrate (i.e., three signs away from the sign where the giver of life was located). Determining which sign or planets were relevant, and how dangerous, demanded all the skill of the experienced astrologer. And the number of years of expected longevity was equal to the number of degrees of right ascension between this place and the aphetic point.

Morandi was presented with a nativity for Urban VIII drawn up for 5 April 1568 at 1:29 pm, with the sun in Aries in the ninth house and Leo in the ascendant.[54] According to the analysis in his letter from "Lyons," the sun was the life-giving planet in this configuration, but its favorable effects were entirely vitiated by the presence of Mars and the moon. Venus, though favorably situated, was no match for these evil planets, especially because the opposition of Saturn in the eighth house in turn blocked it. Urban was very fortunate, said Morandi, to have lived beyond the age of seven.[55] Things started looking better if these directions were progressed to the year 1630. The ascendant would be back, unaccompanied by any unfavorable planets, and Venus would be in its own house and in a good position to counteract Saturn. But none of these benefits were of much use if Urban was doomed to die in any case. That year, a solar eclipse would occur in

June in the sign of Gemini, in the vicinity of Mars, the planetary ruler of late middle age, Urban's current stage of life. Concerning the influence of solar eclipses, Morandi paid less attention to Ptolemy's *Tetrabiblos* than to the pseudo-Ptolemaic *Centiloquium* (which he erroneously attributed to Hermes Trismegistus). Based on this reading, his conclusion was inescapable: Urban VIII was doomed. The fact that Rome was under Taurus, a whole sign away from Gemini, where the eclipse would be occurring (although Italy was sometimes said to be under Libra), was not enough to save his life.

This prophecy did not go unchallenged, even at the monastery. One frequent visitor who objected was Raffaele Visconti, an advisor to the Holy Office. He explained the reasons for his dissent in an anonymous letter that he too pretended to have been sent from Lyons, dated 21 February 1630, in reply to the previous one of Morandi.[56] According to him, Mars in a right quartile with the sun, i.e., four signs away, a disharmonious position tending to accentuate the bad in whatever revolutions were occurring, combined with the moon, the sun's natural antagonist, in almost the same place in the sky, was far more significant than Morandi realized. Rather than simply offending the sun, this double check removed the sun entirely from its position as the giver of life in the chart. Moreover, Visconti found the negative effects of Saturn to be counteracted by the vicinity of the dragon's head and the edge of the house of Virgo, not to mention the antiscia or corresponding point opposite Venus. The progression to the year 1630, he contended, was far more favorable than Morandi made out. Not only would the ascendant be back, unshadowed, and in a good position with respect to Saturn, but Mars too would be in an aspect with Saturn, whose excessive frigidity it would temper still further by its excessively dry effects.

The crucial question, according to Visconti, was whether the solar eclipse of 1630 would do any harm. In fact, two eclipses had previously occurred in the year 1624—one in Urban's ascendant and another in quadrates with the sun and moon in his nativity; and he was never better. Most likely, Visconti suggested, the new eclipse would be harmless unless the pope happened to find himself in a city badly situated with respect to the area of the sky where the eclipse occurred. Thus, if Urban went to where Gemini ruled, he could die; but in Rome he was safe for now. The first signs of real trouble Visconti was able to find only in the years 1643 or 1644, when, the combination of negative planets could well be fatal. The future would prove just how accurate this prognostication actually was—if not in its attribution of the cause of Urban's death, at least concerning the time.

But Visconti's view was drowned out in the universal chorus of assent that greeted Morandi's analysis within the monastery. No one, it turned out, could convince Francesco Usimbardi and two academicians, namely, Vitellio Malespina and Bartolomeo Filicaia, that the prediction conveyed *viva voce* to them by Morandi might not be true.[57] Whereas the poet Francesco Bracciolini believed the prophecy because of the authority of its bearer, the astrologer Francesco Lamponi believed it because he had reached the same conclusions on his own— to the delight of his patron, Cardinal Desiderio Scaglia, whose papal aspirations suddenly seemed less fanciful than ever.[58]

For Morandi, as for Galileo, the chief value of his discovery lay in the number of defenders it might win to his cause—in other words, the cause of astrology. Just in case news about the prophecy did not spread far enough spontaneously, Morandi himself communicated it by mail to Simon Carlo Rondinelli, librarian to Giovanni de' Medici in Florence, and to many other correspondents whose identities he refused to reveal to the court.[59] Soon the news found its way into the manuscript newsletter network; and just as inevitably, it went through a few transformations along the way. With mediocre comprehension, an anonymous newsletter writer on the payroll of the duke of Modena attributed the prophecy to Tommaso Campanella, much to the chagrin of the latter, who immediately rushed to defend his already precarious position at the papal court.[60] Another newsletter, as we said in the last chapter, attributed the prophecy to "the astrologer" Galileo Galilei, who, he said, happened to be in town "trying to publish a work against the Jesuits"—presumably, the *Dialogue Concerning the Two Chief World Systems*. Not satisfied with the prediction about the pope's death, the writer went on, "the astrologer" had apparently added another prediction about the death of Taddeo Barberini, the pope's nephew, not to mention yet another one about the birth of a male heir to Donna Anna Colonna, Taddeo's wife, and about the future peace of Italy.[61] In many forms and in many cities, the prediction made the news, and Naples was by no means the only place besides Rome where it was widely reported as a proven fact.[62]

As the prediction about Urban VIII rapidly slid out of its author's control, it became a part of the bitter factional struggle in Rome. Among those of the pro-Spanish faction who could hardly contain their joy at the prospect of an imminent conclusion to Urban's notorious French favoritism was Ludovico Ridolfi, brother of the outgoing Maestro of the Sacred Palace. He forthwith recruited Raffaele Visconti to help communicate his sentiments "by day to the princes" (so Tommaso Campanella later recalled) and "by night, to the Spanish." With Urban's fate apparently decided, the question arose of who might be the next pope.[63] A flood of clandestine publications concerning the possible outcomes of the next conclave encouraged further speculation.[64] The Spanish cardinals forthwith set sail for the coast of Italy and the German ones began the difficult journey across the Alps in order to be in Rome in case a new conclave was called.[65]

Unfortunately, Morandi's careful calculations were in error, and Urban lived on for another fourteen years, to 1644. He was not killed in 1630. Nor was he amused. Instead of calling in Campanella, who by that time was out of favor for having revealed the embarrassing incantation scene with the pope in one of his myriad publications, Urban decided, so to speak, to shoot the messenger. He ordered the judges to put Morandi in custody and, in effect, to throw the book at him—accusing him of trafficking in prohibited literature, political chicanery, judicial astrology, and, worst of all, the publication and dissemination of astrological information.

Soon afterwards Morandi was found dead in jail—according to the doctor's report, which is in our trial record, due to complications from a high fever and not, the doctor is quick to add, from poisoning. Perhaps the doctor protested too much. A quick dispatching of the suspect may well have been ordered to protect the reputations of the many illustrious persons in the curia, more and more of whose names were being dredged up during the course of the trial and copied down in the record as "Cardinal ———" or "Monsignor ———." The following year Urban published the severest bull yet against the astrologers, emphasizing, with obvious reference to Morandi, the aggravation of the crime when the accused disseminated his astrological conclusions far and wide.[66]

In vain, the astrologers protested that canceling astrology would cancel all astronomy because people mainly looked at the stars to see what they could get out of them. Said one, "I think few people really care about Mars or Jupiter unless they can incorporate this knowledge into a prediction." Without this motive, the astrologer suggested, there would be no stargazing, no planetary calculation, and eventually no means even of determining the proper time for the Easter cycle (based on the vernal equinox and the phases of the moon)—with grave consequences for religion, for faith, and for the Holy Catholic Church.[67] Kepler, echoing an argument Galileo made about astronomy, suggested that prohibiting astrology was tantamount to prohibiting all philosophy, inasmuch as the science of the stars was bound up with every other aspect of natural knowledge. And Campanella reinterpreted the bull in such a fashion that astrologers should have been able to go on much the same as before.[68]

Urban VIII was not deterred. Indeed, for him, the less stargazing the better; and the Galileo case, which came up the year after the publication of the antiastrology bull, only confirmed this view. He was not disposed to permit any more meddling with the stars, in spite of his confidence in the methods of defense suggested by Campanella. But already astrology had passed the apex of its credibility. With predictions having become mere merchandise in the game of political and social favors, nagging suspicions began to emerge. The demarcation between the tradition of sincere scientific astrology of writers like Cardano and Campanella and astrology as a product for consumption became harder and harder to draw. The secrecy and imprecision that in earlier years had been its principle fascination, in an age more and more preoccupied with absolute truths and with accurate measurements, became its chief defect. Gradually, astrology came to be displaced by sciences that could deliver such certainty and precision—at a lower price; and astrology receded into the murky shadows of unorthodoxy and eccentricity.[69]

Over the long run, astrology was defeated by the same techniques of investigation that its practitioners had begun to promote in competition with the astronomers. It was also defeated by the same entrepreneurial behavior that had become prevalent in other fields. The marketplace for natural knowledge that the more sophisticated astrologers had contributed with the astronomers to create, lived on. For serious seekers after natural knowledge, the problem now, as the next chapter shows, was where to go from here.

Notes

1. Morandi's trial is in ASR, Governatore, Processi, sec. XVII, b. 251, henceforth referred to as *Morandi trial*. The bibliography on Morandi is not long: Germana Ernst, "Scienza, astrologia e politica nella Roma barocca," in Eugenio Canone, ed., *Bibliothecae Selectae, da Cusano a Leopardi* (Florence: Olschki, 1993), 217-52; in addition, Torello Sala, *Dizionario storico-biografico di scrittori letterati ed artisti dell'ordine di Vallombrosa*, 2 vol. (Florence: Istituto Gualandi, 1929), 76-77; D. F. Tarani, *L'Ordine Vallombrosano. Note storico-cronologiche* (Florence: Scuola Tipografica Calasanziana, 1920), 127; Antonio Bertolotti, "Giornalisti, astrologi e negromanti in Roma nel secolo XVII," *Rivista europea* 5 (1878): 478-79; Luigi Fiorani, "Astrologi, superstiziosi e devoti nella società romana del Seicento," *Ricerche per la storia religiosa di Roma* 2 (Rome: Edizioni di Storia e Letteratura, 1978), 97ff. In addition, Galileo Galilei, *Edizione nazionale delle opere*, 20 vols. ed. Antonio Favaro, 3rd ed. (Florence: G. Barbera, 1967), 14: 135; Giacinto Gigli, *Diario romano (1608-70)*, ed. G. Ricciotti (Rome: Tumminelli, 1958), 118; Tommaso Campanella, *Lettere*, ed. Vincenzo Spampanato, Scrittori d'Italia, no. 103 (Bari: Laterza, 1927), 287-88. Now see my *The Last Prophecy of Morandi: Making the Occult in Baroque Rome* (Princeton: Forthcoming).
2. In general, on the role of the occult in this period, Brian P. Copenhaver, "Natural Magic, Hermetism and Occultism in Early Modern Science," David C. Lindberg and Robert S. Westman, eds., *Reappraisals of the Scientific Revolution* (Cambridge: Cambridge University Press, 1990). In addition, the articles in Ingrid Merkel and Allen G. Debus, eds., *Hermeticism and the Renaissance. Intellectual History and the Occult in Early Modern Europe* (Washington, D. C.: Folger Books, 1988); Wayne Shumaker, *The Occult Sciences in the Renaissance* (Berkeley: University of California Press, 1979); Brian Vickers, ed., *Occult and Scientific Mentalities in the Renaissance* (Cambridge: Cambridge University Press, 1984); Charles Webster, *From Paracelsus to Newton: Magic and the Making of Modern Science* (Cambridge: Cambridge University Press, 1983); Paola Zambelli, *L'ambigua natura della magia* (Milan: Il Saggiatore, 1991), especially chap. 1; Cesare Vasoli, "Ermeticismo e cabala nel tardo Rinascimento e primo Seicento," in Fabio Troncarelli, ed., *La città dei segreti* (Milan: Angeli, 1989), 103-18; and Vasoli, ed., *Magia e scienza nella civiltà umanistica* (Bologna: Mulino, 1976).
3. Galileo's debate about this with Cesare Cremonini is examined in Luigi Olivieri, *Certezze e gerarchia del sapere* (Padua: Antenore, 1983), 136.
4. Germana Ernst, "Aspetti dell'astrologia e della profezia in Galileo e Campanella," Paolo Galluzzi, ed., *Novità celesti e crisi del sapere. Atti del convegno internazionale di studi galileiani, 1982* (Florence: Giunti-Barbéra, 1984), 255-66.
5. Mary Ellen Bowden, *The Scientific Revolution in Astrology: The English Reformers, 1558-1686*, PhD thesis, Yale, 1974, prologue and chap. 1; S. J. Tester, *A History of Western Astrology* (Bury St. Edmonds: Boydell Press, 1987); Ann Geneva, *Astrology and the Seventeenth-Century Mind. William Lilly and the Language of the Stars* (Manchester: Manchester University Press, 1995).
6. Jean-Yves Grenier, *L'economie d'ancien regime. Un monde de l'échange et d'incertitude* (Paris: Albin Michel, 1996); Renata Ago, "Gerarchi delle merci e meccanismi dello scambio a Roma nel primo Seicento," *Quaderni storici* 33 (1997): 663-83; Biagio Salvemini, "Premessa," in *Ibid.*, 621-31; Arjun Appadurai, *The Social Life of Things. Commodities in Cultural Perspective* (Cambridge: Cambridge University Press, 1986).
7. Concerning the 1600 census, Romeo de Maio, "I modelli culturali della Controriforma: le biblioteche dei conventi italiani alla fine del Cinquecento," in his *Riforme e*

miti nella Chiesa del Cinquecento (Naples: Guida, 1973); M. Dykmans, "Les bibliothèques des religieux en Italia en l'an 1600," *Archivium Historiae Pontificiae* 24 (1986): 385-404; and Antonella Barzazi, "Ordini religiosi e biblioteche a Venezia tra Cinque e Seicento," *Annali dell'Istituto storico italo-Germanico in Trento* 21 (1995): 141-228. On ecclesiastical bibliofilia, compare Danilo Zardin, "*Donna religiosa di rara eccellenza.*" *Prospera Colonna Bascapé: i libri e la cultura nei monasteri milanesi del Cinque e del Seicento* (Florence: Olschki, 1992); and Luciano Allegra, *Ricerche sulla cultura del clero in Piemonte: le biblioteche parrocchiali nell'archidiocesi di Torino, secoli XVII-XVIII* (Turin: Deputazione Subalpina di Storia Patria, 1978).

8. Morandi's lending list is in *Morandi trial*, fols. i-xviii. A list of books attributed personally to Morandi appears in duplicate at fols. 556r-62v. Concerning Roman and Italian libraries in the period, Enzo Bottasso, *Storia della biblioteca in Italia* (Milan: 1984); U. Rozzo, *Biblioteche italiane del Cinquecento tra Riforma e Controriforma* (Udine: 1994); Marino Zorzi, "La circolazione del libro a Venezia nel Cinquecento. Biblioteche private e pubbliche," *Ateneo veneto* 177 (1990): 155-63; Mario Rosa, "I depositi del sapere. Biblioteche, accademie, archivi," in Pietro Rossi, ed., *La memoria del sapere* (Bari: Laterza, 1988), 165-210.

9. Texts referred to, in the most probable editions, include: Aristarchus, *De magnitudinibus et distantiis solis et lunae*, tr. Federico Commandino (Pesaro: Camillo Franceschino, 1572); *Hipparchi Bithyni in Arati et Eudoxi Phaenomena*, ed. Pietro Vettori (Florence: Giunti, 1567); Ptolemy, *Almagestum, Latina lingua donatum*, trans. George of Trebizond (Venice: Giunti: 1527); Copernicus, *De revolutionibus orbium coelestium* (Nuremberg: Petreius, 1543); Galileo Galilei, *Sidereus Nuncius* (Venice: Baglioni, 1610); Johannes Kepler, *De cometis libelli tres* (Augsburg: Aperger, 1619-1620); Idem, *De stella nova* (Prague: P. Sessli, 1606); Idem, *Epitome astronomiae Copernicana*, 7 vols. (Linz: J. Plancus, 1618-22); Erasmus Reinhold, *Tabulae Prutenicae* (Wittemberg: Welack, 1585); Isaac ben Said et al., *Tabulae astronomicae* [i.e., the Alphonsine Tables] (Venice: Hammam, 1492); Johannes Kepler, *Tabulae Rudolphinae* (Gorlitz: Sauri, 1627); Johann Stoeffler, *Ephemeridum . . . a 1532 in alios XX subsequentes* (Tübingen: Morhart, 1531); Giovanni Antonio Magini, *Tabulae generales* (Bologna: Rossi, 1609); and Andrea Argoli, *Ephemerides . . . ab anno . . . 1621 usque ad 1640* (Rome: Facciotti, 1621).

10. For instance, Ptolemy, *Libri quattro*, trans. by Joachim Camerarius et al. (Nuremberg: Petreius, 1535); perhaps also Idem, *Operas quadripartite in platinum sermonic*, trans. Antonius Gogava (Lou vain: Batiks, 1548) or Idem, *Quadripartite indecorum opus*, trans. Johannes Brittuliano (Paris: Porta, 1519); Julius Firmicus Maternus, *De nativitatibus* (Venice: Bevilacqua, 1497); Aratus Solensis, *Phaenomena et prognostica*, either in the version by Philip Melanchthon (Wittenberg: Lotther, 1521), or by Aldus Manutius (Venice: Manutius, 1499); Abenragel, *De iudiciis astrorum*, ed. Bartolomeo de Alten de Nusia (Venice: Sessa, 1503); Henricus Lindhout, *Speculum astrologiae* (Frankfurt: Richter, 1608); Henrik Rantzau; *Tractatus astrologicus de genethliacorum thematum judiciis pro singulis nati accidentibus* (Frankfurt: Wechel, 1593); Rudolph Goclenius, *Uraniae divinatricis* (Marburg: Egenolophi, 1614). In addition, Valentin Naibod, *De annui temporis mensura in directionibus, et de directionibus* (Venice: Damiano Zenari, 1607); Joannes Regiomontanus, *Tabulae directionum profectionumque* (Wittemberg: Welack, 1574).

11. Owen Gingerich, *The Great Copernicus Chase and Other Adventures in Astronomical History* (Cambridge: Cambridge University Press, 1992), title essay; and the articles in Massimo Bucciantini and Maurizio Torrini, eds., *La diffusione del Copernicanesimo in Italia, 1543-1610* (Florence: L.S. Olschki, 1997).

12. Cyprian Leowitz, *De coniunctionibus magnis* (Lavgingae ad Danubium: Salczer, 1564); Philip Melanchthon, *Epistola* contained in *Rudimenta astronomica Alfragrani . . . Albategnius astronomus peritissimus de motu stellarum . . . omnia cum demonstratioibus geometricis et additionibus Ioannis de Regiomonte* (Nuremberg: Petreius, 1537). For Camerarius, see above.

13. *Bullarum, diplomatum et privilegiorum Sanctum Romanorum pontificum Taurinensis Editio* 24 vols., vol. 8, ed. Luigi Bilio (Naples: Camporaso, 1882), 649. For Giambattista Della Porta's defense of astrology, see *Coelestis Physiognonomica* ed. Alfredo Paolella (Naples: Edizioni Scientifiche, 1996). Orig. edition published in 1603. Concerning his system, see the articles in *Giambattista Della Porta nell'Europa del suo tempo* (Naples: Guida, 1990), especially those by Gioacchino Papparelli and Hélène Vedrine; as well as Cosimo Caputo, "La struttura del segno fisiognomico (Giambattista Della Porta e l'universo culturale del Cinquecento," *Il Protagora* 22 (1982): 63-102.

14. For the complex history of Campanella's manuscript on astrology, see the entry by Luigi Firpo in *Dizionario biografico degli Italiani* 17 (1974): 372-401.

15. Keith Thomas, *Religion and the Decline of Magic* (New York: Scribners, 1971), chaps. 1-2. Compare Roy Porter, *Health For Sale. Quackery in England, 1660-1850* (Manchester: Manchester University Press, 1989), chap. 2. In general, for what follows, Gianna Pomata, *La promessa di guarigione. Malati e curatori in antico regime: Bologna, 16-18 sec.* (Bari: Laterza, 1994); as well as Leslie Clarkson, *Death Disease and Famine in Pre-Industrial England* (New York: St. Martin's, 1975).

16. Laurie Nussdorfer, *Civic Politics in the Rome of Urban VIII* (Princeton: Princeton University Press, 1992), chaps. 3, 6; Peter Partner, *The Pope's Men: The Papal Civil Service in the Renaissance* (Oxford: Oxford University Press, 1990); Renata Ago, *Carriere e clientele nella Roma barocca* (Rome: Laterza, 1990), and Maria Antonietta Visceglia, "Burocrazia, mobilità sociale e 'patronage' alla corte di Roma tra 5 e 600: alcuni aspetti del recente dibattito storiografico e prospettive di ricerca," *Roma moderna e contemporanea* 3 (1995): 11-55.

17. Paola Zambelli, "Many Ends of the World. Luca Gaurico Instigator of the Debate in Italy and in Germany," in Zambelli, ed., *"Astrologi Hallucinati": Stars and the End of the World in Luther's Time* (Berlin-New York: de Gruyter, 1986), 239-63. In addition, Robin Bruce Barnes, *Prophecy and Gnosis. Apocalypticism in the Wake of the Lutheran Reformation* (Stanford: Stanford University Press, 1988). Concerning the structure of astrological belief, Franz Böll, et al., *Sternglaube und Sterndeutung. Die Geschichte und das Wesen der Astrologie* (Darmstadt: Wissenschaftliche Buchgesellschaft, 1974), especially chaps. 4-6; but compare Eugenio Garin, *Astrology in the Renaissance: The Zodiac of Life* (London: Routledge, 1983), chap. 1.

18. Germana Ernst, "From the Watery Trigon to the Fiery Trigon. Celestial Signs, Prophecies and History," in Zambelli, ed., *"Astrologi Hallucinati"*, 265-80. In addition, Enrico De Mas, *L'attesa del secolo aureo, 1603-25. Saggio di storia delle idee del secolo XVII* (Florence: Olschki, 1982), especially chap. 1; and Ilan Rachum, "'Revolution' in Seventeenth-Century Astrology," *History of European Ideas* 18 (1994): 869-83.

19. Text reproduced in Mario Pavone, *Introduzione al pensiero di Giambattista Hodierna*, 2 vols. (Modica: Setim, 1981), 1: 153.

20. *Morandi trial*, fol. 183r, deposition of Marcantonio Conti, August 6, 1630.

21. *Morandi trial*, fol. 504r, deposition of Agostino Lamponi, November 5, 1630.

22. *Morandi trial*, fols. IXv and 467r.

23. *Morandi trial*, fol. Vr.

24. Their activities at the monastery are recorded in *Morandi trial*, fols. 467r and 572r (Ghisleri), fols. 130r, 192r, 398r (Scaglia), fols. 110r and 123r (Usimbardi). Scaglia's nativity (for example) is at fol. 1062r.

25. The works in question are Johannes Garcaeus, *Astrologiae methodus, in qua secundum doctrinam Ptolemaei exactissima facillimaque genituras qualescunque indicandi ratio traditur* (Basel: Henricpetrina, 1576); Luca Gaurico, *Tractatus astrologicus*; Girolamo Cardano, *Libelli quinque* (Nuremberg: Petreius, 1547).

26. In his *Uranoscopiae, Chiroscopiae, Metoposcopiae, et Opthalmoscopiae Contemplatio* (Frankfurt: 1608), 272; on which, see Anthony Grafton, *Cardano's Cosmos: The Worlds and Works of a Renaissance Astrologer* (Cambridge, Mass.: Harvard University Press, forthcoming), 267.

27. Kepler, in *Gesammelte Werke*, ed. Franz Hammer et al., 20 vols. (Münich: C. H. Bech, 1937-98), 10: 36; although in his *Tertius interveniens, Gesammelte Werke*, 4: 161, he calls astrology astronomy's "silly little daughter."

28. The quote is in *Morandi trial*, fol. 758r. See also Anthony Grafton, *Cardano's Cosmos*, chap. 3. Concerning the various methods of calculation of astrological charts, see John David North, *Horoscopes and History*, Warburg Institute Surveys, vol. 13 (London: Warburg Institute, 1986).

29. See Gaetano Stano and Francesco Balsimelli, "Un illustre scienziato francescano," *Miscellanea francescana* 43 (1943): 81-149; and the article by G. Odoardi in *Dizionario biografico degli Italiani* 2 (1960): 567-68. Altobelli's chief works were *Tabulae regiae divisionum dodecim partium coeli et syderum obviationum ad mentem Ptolemaei* (Macerata: Bonomi, 1628) and *Demonstratio ostendens artem dirigendi et domificandi Ioannis de Monteregio non concordare cum doctrinam Ptolomaei* (Foligno: 1629).

30. *Advancement of Learning*, ed. Edward Creighton (New York: Colonial Press, 1900), 3: 4 (p. 90 in this edition). Not in the least Whiggish of course is Paolo Rossi, *Dalla magia alla scienza*, 2nd. ed. (Turin: Einaudi, 1974); and especially see D. P. Walker, "Spirits in Francis Bacon," in *Francis Bacon: Terminologia e fortuna nel secolo XVII* (Rome: Edizioni dell'Ateneo, 1984), 315-27. Of course, I am not confusing Bacon's empiricism with the "empirics" whom he condemns. See Didier Deleule, "Experientia-experimentum chez F. Bacon," in *Ibid.*, 59-72.

31. *Morandi trial*, fol. 108r, Morandi deposition, 15 July 1630.

32. Here and below, *Morandi trial*, fols. 959r and ff.

33. See, for instance, Charles B. Schmitt, *The Aristotelian Tradition and Renaissance Universities* (London: Variorum Reprints, 1984).

34. Egidio' translation was first published as *Liber quadripartiti* (Venice: Ratdolt, 1484). Camerarius' translation of Books 1 and 2, with parts of Books 3 and 4, was published as *Libri quattuor* (Nuremberg: Petreius, 1535); Melanchthon's was published by Oporinus at Basel in 1553; Antonius Gogava's was entitled *Operis quadripartiti in latinum sermonem* (Louvain: Batius, 1548). This was the version used by Cardano in his commentary.

35. Compare James Lattis, *Between Copernicus and Galileo: Christopher Clavius and the Collapse of Ptolemaic Cosmology* (Chicago: University of Chicago, 1994), chap. 1. The most comprehensive analysis of Ptolemy's theories and their background is still Auguste Bouché-Leclerq, *L'astrologie grecque* (Paris: 1899; repr. Aachen: Scientia Verlag, 1979).

36. Among other studies on the impact of heliocentrism there is Michel-Pierre Lerner, *Tre saggi sulla cosmologia alla fine del Cinquecento* (Naples: Istituto Italiano per gli Studi Filosofici, 1992).

37. Nifo's work was *De diebus criticis* (Venice: Ponti, 1504). On Girolamo Cardano's contributions concerning critical days, Nancy G. Siraisi, *The Clock and the Mirror. Girolamo Cardano and Renaissance Medicine* (Princeton: Princeton University Press, 1997), chap. 6. Pseudo-Ptolemy's *Centum Ptolomaei sententiae* were published by Aldus Manutius in Venice (1519) among others. Concerning astrology and humoral theory see

Fritz Saxl, Raymond Klibansky, and Erwin Panofsky, *Saturn and Melancholy* (London: Nelson, 1964), part 2, chap. 1.

38. Germana Ernst, "Aspetti dell'astrologia e della profezia in Galileo e Campanella," in *Novità celesti e crisi del sapere. Atti del convegno internazionale di studi galileiani, 1982*, ed. Paolo Galluzzi (Florence: Giunti-Barbéra, 1984), 255-66.

39. Of Offusius, note his *De divina astrorum facultate* (Paris: Royer, 1570). The work in question of Robert Fludd was *Utriusque Cosmi maioris scilicet et minoris Metaphysica, Physica atque Tecnica Historia*, 2 vols., vol. 1 (Oppenheim: Galler, 1617), Treatise 1, Book 2, 45ff and Treatise 1, Book 5, chap. 10, 146ff. A late example of astrological medicine, among many: Giovanni Antonio Magini, *De astrologica ratione ac usu dierum criticorum* (Venice: Zennari, 1607), present in the S. Prassede library. For what precedes, Mary Ellen Bowden, *The Scientific Revolution in Astrology*, chap. 2; Lynn Thorndike, *A History of Magic and Experimental Science*, 8 vols. (New York: Columbia University Press, 1923-58), 6: 113 and 7: 98; on Fludd, Allen G. Debus, *The Chemical Philosophy: Paracelsian Science and Medicine in the Sixteenth and Seventeenth Centuries*, 2 vols. (New York: Science History Publications, 1977); William H. Huffman, *Robert Fludd and the End of the Renaissance* (London: Routledge, 1988).

40. Johannes Kepler, *Mysterium Cosmographicum* (Tübingen: Georgius Gruppenbachius, 1596), chap. 9. In addition, here, Mary Ellen Bowden, *The Scientific Revolution in Astrology*, 112. In addition to Gérard Simon, *Kepler: Astronome, Astrologue* (Paris: Gallimard, 1979), there is Fernand Hallyn, *Structure poétique du monde: Copernic, Kepler* (Paris: Seuil, 1987) and J. V. Field, "A Lutheran Astrologer: Johannes Kepler," *Archive for the History of the Exact Sciences* 31 (1984): 189-272, all to be compared with Bruce Stephenson, *Kepler's Physical Astronomy* (New York: Springer Verlag, 1987).

41. *De fundamentis astrologiae certioribus*, in *Opera omnia*, 1: 417-38, Thesis 4.

42. *Opera omnia*, 5: 285, quoted in Lynn Thorndike, *A History of Magic and Experimental Science*, 7: 22.

43. For the different chapter divisions in the various editions see Wolfgang Hübner's edition, more correctly entitled *Apotelesmatika*, of Ptolemy's *Tetrabiblos* (Stuttgart and Leipzig: G. Teubner, 1998), vii-li.

44. Compare Bouché-Leclerq, *L'astrologie grecque*, 177. For the preceding clause, *Ibid.*, 245.

45. Most probably, the book of Albumasar present in the library was the *Flores Astrologiae* first printed by Sessa in Venice, 1488.

46. *Morandi trial*, fol. 943r.

47. *Edizione nazionale delle opere di Galileo Galilei*, 14: 107, letter dated May 24. In addition, *Morandi trial*, fol. 572r.

48. For the performance aspect of science in this period see Mario Biagioli, *Galileo Courtier* (Chicago: University of Chicago Press, 1993).

49. Nicolò Barozzi and Guglielmo Berchet, *Relazioni*, 211, from the report of Pietro Contarini in 1627.

50. Cases mentioned in this paragraph are documented in Mary Ellen Bowden, *The Scientific Revolution in Astrology*, 110; Anthony Grafton, *Cardano's Cosmos*, chaps. 2, 6.

51. Luigi Amabile, *Tommaso Campanella ne' castelli di Napoli, in Roma ed in Parigi*, 2 vols. (Naples: Morano, 1887), 2: 153; 172ff.

52. Described by Campanella in his *De siderali fato vitandi*, a pirate version, published in 1629 without his knowledge, as Book 7 of his *Astrologicorum* (Leyden: Prost, 1629 [actually, Rome: Brogiotti, 1629]). See Germana Ernst, *Religione, ragione e natura. Ricerche su Tommaso Campanella e il tardo Rinascimento* (Milan: Angeli, 1991), chap. 1.

53. On this subject, Auguste Bouché-Leclerq, *L'astrologie grecque*, 416-28. Ptolemy describes his method in *Tetrabiblos*, ed. F. E. Robbins (Cambridge, Mass: Harvard University Press, 1971), 3: 10.

54. Several versions of Urban VIII's horoscope can be found in *Morandi trial*, with slight differences, at fols. 13r, 1050r, 1264r.

55. No copy of Morandi's document remains. What follows is a reconstruction based on second-hand information presented in *Morandi trial*.

56. Two copies of the document dated "Lyons, 21 January 1630" containing the prophecies regarding Urban VIII exist in *Morandi trial*, both in Morandi's hand: a rough draft at fols. 90r and following; and a fine copy at fols. 486r and following.

57. *Morandi trial*, fol. 110r, Morandi deposition, 15 July 1630.

58. *Morandi trial*, fol. 536r, letter dated 21 March 1628.

59. *Morandi trial*, fol. 110r, Morandi deposition, 15 July 1630.

60. Luigi Amabile, *Tommaso Campanella*, 2: 149, quoting a report in the files of the Este secretary of state, dated 4 May 1630.

61. Galileo Galilei, *Edizione nazionale delle opere*, 14: 103, newsletter dated 18 May 1630.

62. Tommaso Campanella, *Lettere*, ed. Vincenzo Spampanato, Scrittori d'Italia, no. 103 (Bari: Laterza, 1927), 288, letter to Urban VIII dated 9 April 1630.

63. Tommaso Campanella, *Lettere*, 287-88. Cited in Germana Ernst, "Scienza, astrologia e politica nella Roma barocca," in Eugenio Canone, ed., Bibliothecae Selectae, da Cusano a Leopardi (Florence: Olschki, 1993), 221.

64. Luigi Amabile, *Tommaso Campanella*, 2: 150, from a report in the files of the Este secretary of state, dated 18 May 1630.

65. As later recalled by Theodore Ameyden in his manuscript *Elogia*, cited by Alexandro Bastiaanse, *Teodoro Ameyden (1586-1656). Un Neerlandese alla corte di Roma* ('s-Gravenhage: Staatsdrukkerij, 1968), 51.

66. *Bullarum, diplomatum et privilegiorum*, vol. 14, ed. Francisco Gaude (Turin: Dalmazzo, 1868), 211, dated 1 April 1631, and entitled *Inscrutabilis*.

67. *Morandi trial*, fol. 753r.

68. *Atheismus triumphatus* (Paris: Dubray, 1636), special section entitled *Disputatio an bullae SS. Pontificum Sixti V et Urbani VIII . . . calumniam in aliquo patiantur*, 255-73.

69. Compare explanations for the decline of seventeenth-century astrology in Ann Geneva, *Astrology and the Seventeenth-Century Mind: William Lilly and the Language of the Stars* (Manchester: Manchester University Press, 1995), chap. 9; Bernard Capp, *Astrology and the Popular Press: English Almanacs, 1500-1800* (London: Faber and Faber, 1979), 276ff. Interesting comments on the historiography of astrology are in Cesare Vasoli, ed., *Magia e scienza nella civiltà umanistica* (Bologna: Il Mulino, 1976), Introduction.

Chapter 3:
Printing Natural Knowledge

"The purpose of writing may be to persuade the wise and intelligent; or else it may be to persuade everyone in town." So wrote Giovanni Alfonso Borelli, professor at the university of Pisa and already the author of an important work on Euclid, to his ex-colleague Marcello Malpighi, just starting out on the faculty at the university of Bologna and about to publish his fundamental work revealing the microstructure of the lungs. Deciding what to publish, and for whom, was no easy matter, still some thirty years after the twin trials of Orazio Morandi and Galileo Galilei. "But if you choose to write to the common citizens of Bologna to persuade them about the calumnies made against truth," Borelli continued, "I conceive that in this case you may use every effort to deride any person who peddles a pile of idiocies just to offend virtuous and meritorious people." [1] Of course, for Borelli, "virtuous and meritorious people" meant persons who followed empirical method for the study of nature, exemplified by Galileo; and "truth" meant the truth drawn from the use of that method. "Idiocies" meant ideas conceived by those whom the Galileians conceived as dogmatists and mystifiers. Still in the 1660s, whether Galileo's disciples and their followers would triumph against the dogmatists and mystifiers was by no means clear. The story of their campaign to capture public attention, with all its fits and starts, is far less well known than the content of their ideas. It will occupy us here.

The decision to take the campaign for the new science to "the common citizens of Bologna," or Venice, or Florence, or Rome, was not taken lightly. "Learn at Galileo's expense," Michelangelo Ricci told Donato Rossetti, Borelli's student and then a professor at Pisa, "He ran into so much trouble just because he picked fights."[2] He would not go so far as to suggest, with the literary gadfly Gregorio Leti, that "whoever wishes to look minutely into the effects that an author's ideas might have on the minds of people would never write a book."[3] But as Roman correspondent to the discreet Accademia del Cimento in Florence, he well knew how to avoid controversy. And the best way was to ignore the revolution in print and work behind the scenes.

Enthusiasm about publishing became unconditional only around the 1690s. By that time, a host of new genres and methods had emerged, from natural science handbills to periodical publications. As the Italian practitioners of the natural sciences began to understand their market, they experimented with new techniques for appealing to it. To be sure, the boundary between those interested in investigating natural phenomena and those who lived off the investigations of others was not particularly distinct. Nor was there a distinct division between those interested in continuing the innovative tradition of Galileo and those interested in following a more traditional line of research. But enough can be said about the emergence of scientific entrepreneurialism in seventeenth-century publishing in this chapter to serve as a background for our discussion of the education market in the next chapter.

Diffidence about publishing natural knowledge had many causes, and the confusing behavior of the authorities in charge of monitoring the printing press was certainly a major one. Church officials ignored open expressions by influential French clerics like Marin Mersenne and Pierre Gassendi in favor of the same heliocentric views condemned in the trial of Galileo. Yet they seemed to be just as adamant as in Galileo's time in opposing the publication of new ideas about science as late as 1663, when they put Descartes' works on the Index. Then, within four years of the condemnation of Descartes, they permitted open expressions of various versions of the mechanical philosophy plainly in their midst, in the Roman *Giornale de' letterati*, run by Francesco Nazari and collaborators from 1668 ("what would be wrong . . . if the generation and corruption of things should be caused only by local movements of the atoms and there would be no substantial forms in the world"?[4]) Elsewhere in Italy the situation was not much clearer. In Bologna, the local censor provided Malpighi with a perfect pretext for seeking the advantages of an English publisher by objecting to the Protestant addressee of the letter on silkworms (Henry Oldenburg, later changed to the Royal Society) even though he had approved of similar dedications on previous occasions. In Naples, the local Inquisition said nothing about Lucantonio Porzio's openly Cartesian explanation of matter and motion in *Del sorgimento de' licori* (1670) concerning capillary action, in spite of the cries of the Aristotelians, who had enough clout to induce a Cartesian witch hunt later on.[5]

Furthermore, anything printed was not only subject to being read, but also to being stolen. Borelli worried about putting the Accademia del Cimento into direct contact with its sister-academy, the French Académie des Sciences. "It seems to me that we ought to be informed about the activities and speculations of that academy," he noted, "[but] I am hesitant . . . [since] I cannot be sure that those French gentlemen . . . [will not] follow the old custom of making foreigners the authors [of our discoveries]."[6] Printers all over Europe well knew that intellectual property, however fiercely defended by writers, was not yet a legal category.

Others believed the potential audience for natural knowledge in Italy was too small. To secure credit for his discoveries, Malpighi was content to keep his friend and patron Cardinal Scipione Borghese up to date on all his activities at the University of Bologna; but instead of publishing his letter on silkworms in Italy, he published it in London, and in the one language all the experts understood: Latin. His correspondents agreed. "[Your book] will have far greater success," noted his lifelong friend, the Calabrian physician Giambattista Capucci, "than it would have had in Venice or Bologna . . . and it will have a wider and longer itinerary than in our cities."[7]

Indeed, in Italy as elsewhere, non-print forms of communication continued to be the fundamental vehicle of scientific ideas. Letters possessed the advantage of immunity to the authorities that monitored the press, and they were never seized in customs raids along with books and journals. They usually took two to four weeks to cross Europe, as against the several months to a year required for news to appear in the journals.[8] Even though journals could promise wider diffusion, correspondence habits established in the late Renaissance, by which busy virtuosi received and answered as many as a dozen letters a day, guaranteed that information got out when they wanted it to. And those virtuosi who did not want to bother to spread their ideas themselves could expect professional letter-writers, successors to Marin Mersenne and Nicolas Claude Fabri de Peiresc, like the Royal Society secretary, Oldenburg, and the grand ducal librarian in Florence, Antonio Magliabechi, to do it for them. Magliabechi accordingly justified his services to Malpighi by pointing out how "the time you have for writing your very learned works is so precious." His correspondence, amounting to over 20,000 entries, became the most voluminous of them all.[9]

Manuscripts continued to be a legitimate form of publication in Italy and abroad. Borelli's accounts of the Accademia del Cimento's solution to the Saturn ring controversy, passed around among members of the academy and its correspondents in Rome and The Hague, was considered a sufficient solution to the quarrel raging between Christiaan Huygens and Honoré Fabri.[10] Indeed, manuscripts still had considerable advantages over printed publications. Malpighi could continue to put changes on his extended letter on conglobate glands while handing the manuscript around to friends. As soon as he tried to print it his troubles began. With so many manuscripts of the letter in circulation, his printer managed to get hold of the penultimate—instead of the final—version; so Malpighi was forced to have his latest corrections penned into all the printed copies by hand.[11]

Finally, personal encounter as a means of exchanging information continued to be just as important an alternative to print in the late seventeenth century as it had been when Galileo packed his telescope and went to Rome to demonstrate the satellites of Jupiter. Into Italy came the English physician John Ray, the Danish naturalist Nicolas Steno, and the French astronomer Adrien Auzout. Out went Lorenzo Magalotti, ex-secretary of the Accademia del Cimento (to England) and Bolognese university astronomer Giovanni Domenico Cassini (to become royal astronomer in France). To make sure sophisticated travelers missed nothing, special guidebooks began to appear, such as Gregorio Leti's *Italia regnante* and Jean Huguetan's *Voyage en Italie*. "The collection of Count Mascardi [in Verona]," said a typical passage in the latter, "is full of natural rarities, antiquities, and good paintings." And he provided a list of all the "learned authors and curious and ingenious artisans of Italy" who could be called upon for consultation.[12]

At least until the 1670s, Italian natural science investigators mostly put the finishing touches on the communications mechanisms they had inherited from their sixteenth-century predecessors, without trying to modify them too much. What Galileo had so innovatively put together, namely, the two main publics of experts and amateurs, they solemnly put asunder. Let us examine some of their productions.

In disputes with their colleagues, the practitioners perfected the standard academic point-by-point refutation. It was not very effective unless readers knew something about the works opposed. Since specialized treatises had very small press runs, a disputant usually fit long segments of the disputed work into his own treatise, as did Galileo's disciple Carlo Dati, the erudite librarian of Cardinal Giulio de' Medici, in refuting an attack on his fellow-Galilean Evangelista Torricelli. Unfortunately, such refutations only appealed to readers directly interested in the subject at hand and left everyone else completely cold. "Indifferent readers often seem disgusted," explained Borelli to Malpighi, "by those semantic quarrels and barbs, since they do not feel the same passion [as the disputants]."[13] A few practitioners accordingly combined the academic refutation with the dialogue. This allowed them at least some of the advantages of a form that was still widely regarded as "looser and freer" than other forms, without the difficulties, which they all recognized, of writing a pure dialogue.[14] An example was the refutation by Stefano Degli Angeli, an assistant to the renowned mathematician Bonaventura Cavalieri at Bologna and later professor at Padua, of the geocentric arguments of the Galileans' archenemy, the Ferrarese Jesuit virtuoso Giovanni Battista Riccioli.[15] Unlike Galileo, he made no concessions to readers who might have difficulty following his explanations. Nevertheless, a dialogue seemed less dull than the usual approach, even where the amusing figure of Simplicio, invented by Galileo to personify dogmatism, was replaced by pedantic citations in the original Latin from Riccioli's treatise.

The one figure that seemed to carry forth Galileo's campaign for popular science was Alessandro Marchetti. The mid-century quarrel concerned atomism, and Marchetti, a mathematician at Pisa, believed the public ought finally to be brought up to date. "[The doctrine] does not deserve to be hidden indoors," he wrote to Leopoldo de' Medici, "but on the contrary, it ought to be propagated and taught to all men."[16] And by providing the first Italian paraphrase of Lucretius' *De natura rerum* he hoped to do just that. To make sure none of his readers would mistake his paraphrase for a mere poetical diversion or its unusual doctrines for mere fantasies designed to amaze the reader, he changed the geographical and chronological setting of the poem from ancient Greece to modern Europe. In the passages on Sicily, the country of Empedocles, he inserted a eulogy of Giovanni Alfonso Borelli; and in those on the birth and death of the world, he inserted one on Pierre Gassendi. Furthermore, to make sure readers would not pass it off as just another translation, he inserted an elaborate disclaimer of the famous doctrines on the mortality of the soul in Book Three, like the anti-heliocentric letter prefaced by Andreas Osiander to Copernicus' *On the Revolutions of the Heavenly Spheres*. "So much do I detest [these views]," he protested, "that I shall be ready to defend their opposite, whenever necessity calls, not only by employing all my intellect and strength but also by spilling all of my blood."[17] Few readers could have resisted the temptation to read on.

But poetry was a dangerous medium, as Marchetti soon found out. It seemed to savor too much of Galileo's animated campaigns for awakening interest in science in "the whole city" that might corrupt the minds of the excessively excitable Florentines.[18] Officials in Florence, like their counterparts all over Italy, agreed that transmitting the latest ideas in accessible form was worse than transmitting them in complex and uninviting treatises.[19] So the latter were to be permitted, and the former were not. Still, no one could be sure. Leopoldo de' Medici accordingly prohibited the printing of Marchetti's work in Florence while praising it as a monument to Tuscan scholarship. "He wanted an autograph copy for himself," Marchetti recalled, "and another for the grand duke."[20] Marchetti meanwhile was assured of full credit, as the work was referred to in book lists sent to France as though it were a published book.

Most of the Italian practitioners preferred to hold back on printing for nonexpert audiences and divulge their ideas orally to a single, well-defined and relatively small fragment of the general audience—namely, students in the university medical faculties where most of them taught.[21] Preparing scores of faithful students who could later lend authoritative support to one or another side of a scientific dispute seemed like a more secure method than either trying to win popular applause or trying to convert obstinate professors, which had been Galileo's two chief tactics. "According to the order of nature," Borelli remarked, such students "will have to remain in the world after the old intractable ones are dead."[22] This was, he added, "a path that leads pleasantly to our goal without damaging our allies." Lorenzo Bellini, professor at Pisa and one of the best-known anatomists of his time, agreed. "It is amusing to see the university every year more filled with my protégés," he remarked, "and now the principal chairs of practical and theoretical medicine are occupied with much applause by my students, among whom are numbered also the [professor of] simples, and various logicians and philosophers."[23] Best of all, practitioners could thus avoid any of the difficulties posed by publishing their ideas in printed form—including the difficulty of having to prove all the offhand statements they made during the course of extemporaneous in-class delivery. The only disadvantage to this approach, as we shall see in the next chapter, was that the university medical faculties were still not chiefly conceived as places for the inculcation of the latest scientific ideas.

Italian practitioners between the age of Galileo and the 1670s never tried to put all of their less practical ideas together into easily readable prose, lively and with plenty of rhetorical pyrotechnics, a kind of experimental philosophy *pour les dames* calculated to please at least the more adventurous among the general readers.[24] Yet, to capture the attention of busy amateurs more accustomed to baroque novels and works of history than to the abstruse problems of science, practitioners could not stop at proving that the sun was the center of the universe. They also had to say something about how the universe originated, why rain fell, what meteors were made of, what was a rainbow, and what man's position was in relation to God—just the sorts of things that could provide good material for conversation, and that Descartes furnished to French readers with enormous success in his *Dioptrique* and *Météores*.[25]

The one Italian practitioner who came closest to serving such a public was Francesco Eschinardi, mathematics professor at Perugia and later a member of the scientific academy of Giovanni Giusto Ciampini in Rome. He published a manual that avoided the difficulties of synthesis while ostensibly offering what anyone might want to know about the latest developments. "The only aim of this work is to satisfy those who desire to know about important and interesting physicomathematical questions," he claimed, "without having to worry about difficult propositions and long tedious disputations."[26] He assured the reader that he would include only such issues as his readers were likely to find interesting. And accordingly, he briefly explained how sunspots proved the sun's rotation and the satellites of Jupiter proved a fluid, sphereless sky; and he gave the barest outlines of a quantitative physics by way of Galileo's explanations of falling and floating bodies. But after he had omitted all the most complicated discoveries and the most difficult points, he could think of nothing for the second half of his book except to demonstrate a few of the technological achievements that were the byproduct of recent scientific research in the most superficial way possible— various types of clocks, fountains, and hydraulic organs, with a few instructions on how to build them. So cool was the reception to the work that he gave up proceeding to the subsequent volumes of what was supposed to be a multivolume series.

Italian practitioners also missed the opportunity to compose basic textbooks for their less adventurous but far more numerous general readers in public schools all over Italy. The Italianate Frenchman Claude Bérigard, professor at the universities first of Pisa and then of Padua, came closest, when he aimed his *Circulus pisanus* at those who "know little philosophy" and "less Latin"; but as it turned out, the work required a good measure of both. He explained such novelties as the circulation of the blood, Galileo's law of falling bodies, and the corpuscular analysis of matter, and he tried to lighten up the discussion by putting it in the form of a realistic dialogue between Charilaeus the Aristotelian and Aristaeus the modern. Yet instead of organizing his argument by subjects in order to guide the attention of his readers, he followed the organization of his university lectures according to the various books of Aristotle's *Physics*. In addition, he followed the typical post-Renaissance lecture practice of omitting long citations from the parts of Aristotle's original text to which he continually referred. At the university, he would have been correct in assuming that students already had one of the innumerable printed versions in circulation, without which his comments made little sense. Not so among the general public. Readers who did not possess a copy of the *Physics* or a rudimentary knowledge of it, were unlikely to appreciate his attempt to put together an alternative philosophical system to Aristotle's from the various others available in antiquity. To these readers, the work probably seemed like nothing but a disjointed collection of cantankerous cavils. He himself admitted to having "[chosen] the opinions that were the most coherent in order to put a doctrine together" for no other purpose than so it could "stand up to assaults of Aristotle and dare to attack the citadel of the peripatetics."[27] As a result of his uncompromising choices, his *Circulus pisanus* eventually took its place not beside Descartes' *Dioptrique* and *Météores* but beside Pierre Gassendi's complicated avant-garde and equally uncompromising *Syntagma*, among the works that other practitioners like Borelli adopted as textbooks for a university student elite.[28]

While the Italian practitioners of natural science investigations renounced the job of finding agreeable vehicles for their ideas, readers in search of popular science were forced to rely on the widely circulating works of writers at the intellectual fringes. Almanacs and popular astrological works provided attractive confirmations of popular ideas as well as useful advice on everything from the proper time for planting to the proper time for conceiving children. Giovanni Serpetro's *Marketplace of Marvels* offered a fascinating compendium of everything the reader could want to know— from "the virtues of plants and metals," "the nature of animals," and "the anatomy of Man," to the ways, ostensibly as mysterious as their descriptions, of "prolong[ing] life, chang[ing] the complexion of man, loosen[ing] from certain people the force of the imagination, mitigat[ing] the torments and accelerat[ing] the time of clarification." All this it promised to do through "a few general precepts." After all, it noted, "the current century . . . has little inclination . . . toward the arduous and much toward the amusing."[29] However, lack of contact with the scientific debates of his time meant that Serpetro had to rely on such outdated sixteenth-century sources as Ulisse Aldrovandi and Konrad Gesner for whatever he could not find in Pliny, and, for his marvelous cures, on uncritical popular collections like Timoteo Rossello's *De' secreti universali* (Venice: 1559).

The job of diffusing, to the slightly more sophisticated audience of students in the public schools, something closer to what the natural science practitioners were doing, devolved on the rank and file of religious orders. These writers were not mainly interested in winning a place in literature—not even among such scientific greats as the Jesuit virtuosi Giovanni Battista Riccioli and Giuseppe Ferroni.[30] Much less were they interested in converting readers over to their particular scientific views. It was better to teach methodically-ordered traditional positions even if erroneous or questionable, they agreed, rather than still unauthorized theories, even if true. Their modern counterparts have agreed.[31]

Particularly successful at selling traditional science in an attractive package was Giovanni Battista Giattini, teacher at the Jesuit Collegio Romano. He designed his *Physica* for use in the general philosophy course where science was usually situated. The Renaissance textbook format furnished him with a basic model, but he fit together all of Aristotle's main points on given subjects under separate subheadings so readers would never have to turn to the Aristotelian text at all. To make consultation even easier, he offered a handy index indicating the main things to remember concerning each subject—for example, that "matter" is "not a simple body," is "a cause and not just a condition with respect to form" and "has an innate tendency toward form." Meanwhile, he did his best to introduce the fashionable and apparently useful modern empirical scientific approach without adopting the modern conclusions. He described the Torricelli barometric experiment, trying to explain the rising mercury without reference to the void, inadmissible in Aristotelian physics.[32] In his section on cosmology, he showed how to use commonplace analogies, Galileo-style, to clarify complicated problems.[33] Still, science and its methods were not Giattini's main interest; much more important was philosophical correctness. His analysis of recent astronomical discoveries was exemplary. No one could deny that Galileo's responses to the objections against Copernicanism were good ones, he claimed. However, "we must interpret the Scripture literally as much as possible," he insisted; so "I tell you that such a system in any case does not exist."[34] Fortunately, Tycho Brahe's system with the earth at the center and all the planets circling around a revolving sun seemed to fit the appearances almost as well as Copernicus, and the textbook or classroom were not the places to quarrel about the precise interpretation of the best observations. After all, God could decide on whatever system he wanted, and it was impossible to ask his motives for having chosen a geocentric one. Wherever possible, Giattini used science as an encouragement for virtuous civic action. To this the discussion of comets lent itself best. Everyone knew that all motions of planets occurred within the eight perfect spheres of the heavens and the tiny spheres embedded in these spheres within which the planets moved backward in order produce the apparent effect of retrograde motion. But in order to account for the eccentric motion of the comets, the perfect spheres of the heavens would have to be multiplied ad infinitum. Just as a prince, then, occasionally must enact unprecedented laws for the common good, so God, the prince of the heavens, ordains extraordinary things in the sky; and this should keep us mindful that he is our Lord, just as the action of the prince should keep us mindful that we owe him political obedience.[35]

No one could deny that the religious textbook writers succeeded in providing a form of science that could be imparted effectively to the general public. Donato Rossetti observed that the students formed by way of the textbooks, at least in Turin, were the only intellectuals around. "There is no one here who knows how to talk about anything except war, hunting, and building," he wrote, "apart from the Jesuits, who with their usual techniques keep an incredible number of logicians, physicians and metaphysicians."[36] Even Malpighi had to admit that the textbooks had "made letters popular everywhere," although he stiffly added, "they have degraded them."[37]

When periodical journalism first emerged in Italy with the publication of the Roman *Giornale de' letterati* in 1668, the Italian practitioners did not immediately see its potential for putting the job of scientific popularization back into their own hands. They did not even see its promise for putting scientific communication in Italy on a more organized footing. Cultural isolation did not seem to be a problem except in a few outposts like Sicily, home of the Sicilian natural philosopher Paolo Boccone, or Turin, home of Rossetti.[38] The decentralized printing industry, located in no less than 160 Italian cities, seemed to be sufficiently covered by the existing epistolary networks.[39] Works published in out-of-the way places could be procured by letters to friends, as Lorenzo Bellini discovered when he had to find a copy of a nearly untraceable book published by Montanari in Bologna.[40]

The journalists were not much help. The Roman *Giornale* included mostly translations of articles cribbed from two other recently established journals, the *Journal des sçavans* and the *Philosophical Transactions* of the Royal Society. However, the rules of the genre were still in flux.[41] Lack of knowledge about the possible public for journalism called for keeping costs to a minimum, so individual issues of the first journals were rarely more than four to eight pages long. This permitted little space for printing letters or anything else from contributors in full. Hence, most articles consisted of second-hand reports on experiments or on observations done by others. The few original articles were included with a kind of apology. "In order to make the invention [in the book under review] easier for artisans to carry out," the *Giornale de' letterati* explained, "I have thought it a good idea to present [a sample]"[42] In another case, "It was thought well to insert these [mathematical] problems exactly as they arrived from France."[43] The one full-length article in the entire *Giornale* was a six-page description of a bullhorn invented in London by Samuel Moreland.[44]

Practitioners' attitudes soon began to change. As more and more periodical publications containing articles concerning natural knowledge began arriving in Italy from abroad, rapid publicizing of results began to be established as the ideal. Lorenzo Bellini found publication to be the best hedge against contests of priority in discovery. "Publish vigorously!" he told Malpighi. And again, "How long [does your work] have to stay buried? . . . If you do not let me print it, I will do so anyway in spite of you." With so much pressure to print, few virtuosi dared to leave unpublished observations "to the vicissitudes of fortune."[45] Malpighi quickly became a convinced proselytizer for this view. Publication, he said, would "guarantee" Bolognese virtuoso Luigi Ferdinando Marsili's priority for "the invention" of a work he had in progress on the history of engineering.[46] Giorgio Baglivi in Rome called for insertion of an honorable mention of himself in connection with a work he had published six years before, and his only apparent reason was that he had not yet appeared in the journal in question.[47] By that time, virtually every practitioner of note in Italy had been recognized in this way.

Some practitioners who still followed the traditional practice of occasional publication now felt constrained to apologize. "I deposited a message [about my new surveying device] along with other studies of mine with the Most Illustrious Bolognese Senate [sixteen years ago]," explained Geminiano Montanari in the preface to one book, "but I did not publish them until six years ago because I was waiting to put them into a more voluminous work."[48] On another occasion, his student explained that the master had good reason for being so slow to publish. "[Even though] he passed around a few copies to some friends, he held back to make sure that he was on the right track."[49]

Others invented new forms of publication to transmit new categories of information particular to seventeenth-century natural knowledge. One new form was the brief "Observations" favored by Francesco Redi. Yet another was the scientific broadsheet, pioneered by the astronomer Giovanni Domenico Cassini. In a single year, 1666, he published three. Pressed by time and money, he omitted all the rhetorical amenities of the standard printed letter and included nothing but a detailed description of his telescopic discoveries using Eustachio Divini's new 45-palm telescope and, in two cases, diagrams of what he saw, folded over in quarto and printed on both sides with a title at the top, *Martis circa axem proprium revolubilis observationes Bononiae*. And no sooner had he published the first of them than another observer, one Salvatore Serra, tried to claim the discoveries for himself; but his far longer and more pedantic pamphlet, with none of Cassini's attractive diagrams, was doomed to provoke nothing but a few yawns.[50] Redi published his *Osservazioni* and *Esperienze* together in a single twenty-three-page pamphlet in the same year in which he made them (1673), along with a letter about himself written by an English correspondent of Henry Oldenburg. He then republished them a few years later along with a previously independent four-page letter on salt in a new collection called *Opuscoli vari*.

By the end of the century, a new library of publication types concerned with natural knowledge had practically replaced the sixteenth-century genres. Paolo

Boccone published two "Museums" containing various "observations" on everything from newly discovered plant species to the Sicilian earthquake of 1693. The idea was similar to the collection of scientific correspondence, such as that of Leonardo Fioravanti in the sixteenth century and Stanislas Lubienicki in the seventeenth. The difference was that each four- to fifteen-page "Observation," while still addressed to someone, contained none of the amenities of the letter as stipulated by the *ars dictaminis* tradition; instead it began with a preface explaining the truth-value of the evidence presented. "In regard to this tragic earthquake," began one, "I am obliged to rely on the testimonies of the people who were there at the time," he explained; "Your Excellency must therefore . . . bear with me in certain inconsistencies . . . since the same persons were worried and stupefied by terror." In other "Observations," all form disappeared and simple lists of phenomena were allowed to speak for themselves. "Augusta, demesnial city, had a population of 6173," began another entry, "destroyed entirely. . . . The dead: 2300."[51]

As rapid publication became the norm, the many new forms of communication created by natural science investigators filled in whatever gaps remained between brief letters and observations, on the one hand, and, on the other, sustained academic treatises like Borelli's *De motu animalium*. Somewhere in the middle were the new forms designed to present prolonged series of experiments on related problems. Lorenzo Magalotti and his collaborators at the Accademia del Cimento produced the collection of experiments called *Saggi* or "attempts," containing bare summaries of observations with descriptions of the instruments used, without any attempt to situate them into a larger context. The experiments on the barometer said nothing about whether the space above the mercury was a vacuum or extremely rare aether. This form was perfect for academies like the Cimento, devoted to natural knowledge, where the diversity of opinions—in this case, concerning whether the void existed or not and whether atomism was an acceptable description of matter—was so great that the experiments would have taken forever to appear if their editor had waited around for members to agree on what they meant.[52] More appropriate for a single researcher, on the other hand, was the collection of experiments followed by their explanations and called *Pensieri* or "thoughts." This apparently introspective form left Montanari free to explain experiments similar to those of the Cimento on the basis of air pressure, the void, and the corpuscular theory.[53]

When the Venetian printer Girolamo Albrizzi opened the doors of his *Galleria di Minerva* journal in 1696 to original articles untouched by any editorial intervention, he was deluged with more material than he could handle.[54] Naturally, contributors felt free to treat journal articles like any other form, to be combined and exchanged whenever the occasion permitted. Giovanni Domenico Cassini and Geminiano Montanari republished their journal communications independently after the first appearance. Cassini also operated the other way around, republishing his observations on Mars later in the *Galleria di Minerva*.[55]

As the communication revolution got under way, journal publications served many uses—not all of them corresponding to the familiar modern ones.[56] For one thing, they served to show the members of the Galilean school that Italian science was no longer their private affair in a one-sided battle against their Aristotelian and vitalist adversaries. Donato Rossetti combined Galileism with vitalism, and Borelli combined it with atomism, in spite of Galileo's reluctance to accept either of these views.[57] No longer was it possible to presume absolute consensus on, for instance, "the silliness in the philosophy of" the followers of the sixteenth-century vitalist chemist Theophrastus Paracelsus."[58] For Lionardo Di Capua, Elia Astorini, and Sebastiano Bartoli, all reputedly up-to-date natural science investigators of the Neapolitan school, extolled its benefits.[59] Cartesianism was no longer ridiculed; and Giuseppe Valletta, another Neapolitan, insisted on using it as a working hypothesis for his own work. To transform the increasing differences between the Italian practitioners into veritable separate schools, many of them prepared sturdy groups of disciples in the major cities and universities all over the peninsula.[60] All these transformations began to be more apparent as reviews of Italian books began to occupy a larger and larger proportion of the new journals' contents.

Benedetto Bacchini, antiquary and author of the *Giornale de' letterati* in Parma from 1686 eventually gave up reviewing non-Italian books entirely. To be sure, his accurate year-by-year representation of the Italian publishing scene was more useful than what he actually had to say about the books. He believed that Galileism, Cartesianism, Paracelsianism, and everything else were unimportant as long as the practitioners tried to show the connections between their own fields and every other field in the encyclopedia of human knowledge. He tried to provide an example by making the connections between the disciplines of the books he reviewed and whatever other ones seemed closest, and he praised works for the breadth of their synthesis of information of the most disparate kinds. Like the Roman journalists, he hoped his readers would adopt his ideal, the only one for a real "man of letters." In the end, however, he not only failed to establish the connections he sought, but he demonstrated the irreconcilable divergences between various theories concerning basic matters of natural history and erudition and the inability of a single person to come to grips with all of them. And when he admitted that the enterprise "cannot be carried out well," he simply added to the discredit that the most extravagant late seventeenth-century encyclopedists had already brought upon themselves.[61]

Left to find their own way through a scientific maze made more and more complex as the journals continued to keep better track of it, still in doubt about exactly how science ought to be defined, and lacking the rudiments of a professional organization of science, the Italian practitioners eventually reached a tacit agreement about what science ought not to be. One element that at least some of them agreed to reject was just what Bacchini had advocated: encyclopedism. Michelangelo Fardella, professor at the University of Padua and a collaborator on the Venetian *Galleria di Minerva*, was among the most outspoken critics. "Since the mind cannot process and digest such a multitude of diverse and disjointed information," he warned, "it inevitably vomits it up and rejects the major part of it instead of growing and becoming stronger. This is why," he added, "among the men of letters the most disgusting and least knowledgeable are these encyclopedists."[62] His call for more concentrated attention to specialized subjects, in an article which he prominently published in an early number of the *Galleria* the same year as Bacchini's last, was sure to have a powerful effect.

Even more of the Italian practitioners, overwhelmed by the outpourings of their fellows all over the peninsula, agreed on another element in the definition of science: that it should concentrate on empirical observation and leave systematic explanation to the metaphysicians. A comment of Bacchini's concerning specialized scientific works seemed to hit the mark. Observation was the only aspect of science that had been constantly improving over the centuries, he contended. "The best result of so many physical and medical meditations has to do strictly with [the] rational experience [of Man], which has . . . always been constantly true." Attempts at explanation based on true observations, he believed, so far had not produced particularly brilliant results. "Of modern works, the only ones that reach perfection are those that pertain to anatomical inspection and the little that has been gleaned from practical observation."[63] The practitioners did not accept Bacchini's condemnation of all but anatomical observation, but they agreed in establishing strict empiricism as a standard for judging each other and their transalpine counterparts. While Lionardo Di Capua used this criterion to judge all the physiologists of his time, including his favorite, Paracelsus, Vallisneri used it as a basis for his criticism of all biologists from Pliny onward and for his proposal of Malpighi as a methodological model. And the same criterion used against Giacinto Cestoni by Giambattista Trionfetti was also used by Giambattista Capucci against the Royal Society's plan for a natural history of the whole world: namely, that observations must be carefully sifted, controlled, and, most of all, repeated. And indeed, it was just this methodological issue that came to the forefront of Italian debates on embryology in the last years of the century, those debates in which Italian contributions finally placed Italy back at the center of European science.[64]

Toward the end of the century a few natural science investigators in Naples sought a new context for scientific observations upon which all practitioners could agree; and they called it the "Italian philosophy." A background for such an indigenous philosophy could be found in Democritus, they believed; and remnants of it could be traced through all the greats from Galileo to Borelli. First suggested by Giuseppe Valletta, Francesco d'Andrea, and Tommaso Cornelio, founding members of the Accademia degli Investiganti, the idea was taken up again, in a different context, by Giambattista Vico.[65] But by that time natural science investigators had already adopted a new program and broadcast it in the *Giornale de' letterati d'Italia.*[66]

As the novelty of journals began to wear off, practitioners began exploring the various other new uses to which communications to them might be put. One new use was to establish international scientific relations. Publication in the journals of articles by foreigners was a far more evident and effective demonstration of international collaboration than was the exchange of personal letters. This was as true in transalpine Europe as it was in Italy. Information in the *Philosophical Transactions* about Malpighi's work showed the success of the Royal Society's plan of contacting experts all over the world. Information in the *Journal des sçavans* about Giovanni Domenico Cassini's work showed the effectiveness of Louis XIV's policies in bringing together a cosmopolitan group in the Académie des Sciences. Information in the Roman *Giornale* about the work of Adrien Azout showed that these connections were reciprocated by the

Académie's new sister society, Giovanni Giusto Ciampini's Accademia Fisico Matematica in Rome. Another new use of journal communications was to create foreign audiences for otherwise obscure local publications. One example of success in this was Johann Jacob Heinrich of Strasburg, who managed to find out about Italian publications that caused even his Italian correspondents themselves to scratch their heads in despair at ever being able to find them. Similarly, William Waller used knowledge from journals to contradict Malpighi's affirmation that "the study of letters is silent among us [in Italy]" by responding with a list of Italians whose works he wanted to know about.[67] Another new use of journal communications was to provide a forum for collaboration on projects in which virtuosi in various parts of Italy were involved. It was through the Roman *Giornale*, for instance, that various observers all over the peninsula were able to compare notes on the lunar eclipse of 1674.[68]

Yet another use, at least according to astronomer John Flamsteed, was to protect discoveries.[69] No longer did convergence on particular problems and the absence of copyright conventions have to present the threat of nasty priority disputes of the sort that had plagued Galileo and Torricelli. Christiaan Huygens accordingly published his rules of impact and quantity of motion in the *Journal des sçavans* as well as in the *Giornale* of Rome to ensure proper credit to himself. "Since there are similar rules in the last volume of the English journal," he had the Roman journalists explain, "the author declares that he gave his to the Royal Company [sic! for Royal Society] before those were printed in order not to be accused of having stolen from others."[70]

Meanwhile, practitioners discovered the most important use of all for publications: as a new technique for social advancement. Whether at court or among patrician intellectuals or in academies or in educational institutions, ecclesiastical and lay, traditional means of advancement involved winning the esteem of the great through networking, exchanging favors, and engaging in ceremonious practices like public disputations, dinners, and soirées. Publishing, the virtuosi seemed to believe, might allow them to short-circuit the system by appealing beyond these milieus to the wider constituency of public opinion. The new genres of publication were aimed not just at the inbred audience of the virtuosi themselves, nor at the increasingly large crowds of medical students who thronged the halls of the recently refurbished and in some cases recently established universities. They were aimed at reaching a broad range of artisans, merchants, and amateurs of all ranks—"everybody in town," as Borelli said to Malpighi. Said Albrizzi, introducing the *Galleria di Minerva*, "there must be delight even for the unlearned, as well as for every sort of learned." And another journalist addressed his work to "those who are rather uncultivated and would like to get rid of their rusticity."[71]

A widely perceived rebirth of public interest in knowledge encouraged practitioners to address their ideas to a broader audience. There was scarcely a successful tradesman or merchant who did not show off his collection, however small, of carefully preserved books.[72] Whether they understood them or not, all categories of readers, from nobles to clergymen to lawyers, began buying books on "erudite" or "scientific" subjects. A Venetian merchant named Giovanni Tibelli made the collection of scientific texts a veritable hobby.[73] Another named

Gasparo Chechel specialized in astronomy.[74] A painter named Pietro Liberi presented a veritable library of tracts by German chemical philosophers.[75] A lawyer named Pietro Paolo Alberigo possessed rare mathematical works.[76] Even the pastor of the parish of San Giovanni Grisostomo, Giovanni Maria Spadon, possessed not only the works of scientific classics, but also recent textbooks.[77] While science was becoming a method of advancement for university professors, interest in it was becoming a status symbol among the general public.

Not surprisingly, when Albrizzi opened the doors of his *Galleria di Minerva* to whoever would contribute, the number of dilettante readers who sought the prestige of a place in the press was more than anyone could have imagined. Before long, contributors began to act as though progress in science was everyone's affair. One Giovanni Taddei claimed to have demonstrated that Boyle's notion of the elasticity and pressure of the air and not the Aristotelian horror of the vacuum made flesh rise in a cupping glass. And even if contributors' claims were sometimes difficult to accept—such as when a certain Sigismondo Valiano in Siena claimed to have disproved spontaneous generation in three pages—they nonetheless testified to the growing interest in the increase of cultural accomplishments.[78]

In response to an increased demand, practitioners set out to take control of the public view of themselves. For one thing, they sought to replace the popular view of naturalist as a sort of magician with the new one of the naturalist as a source of answers to popular questions about commonplace occurrences. They got a head start from the necessity to explore fields that came directly into contact with the daily experience of their fellow-citizens that had recently emerged in the continuing search for new sources of empirical evidence to support one or another of the various competing scientific syntheses. Their titles in these fields were enough to catch the attention of curious browsers: *Observations concerning those glass drops and threads that completely shatter when broken in any part* was one; *Experiments done by Francesco Redi . . . concerning that water which, it is said, stanches all spilling blood* was another.[79]

Giuseppe Del Papa, a professor at the University of Pisa, was among the first to turn the new interest in commonplace occurrences into an occasion for directly addressing a popular audience in *The Nature of Humid and Dry*. "[Our] natural desire for knowledge," he began, "turns [our] attention only to those things that are very far from our nature and constitution or very rarely present themselves to our senses."[80] Nevertheless, "many . . . things" that we know little about "are close to our essence and very familiar, and . . . are always obvious and exposed to our senses." Such phenomena were no less interesting for their ubiquitousness; indeed, the emptiness of the popular explanations showed that something much more complex must be going on than appeared at first sight. To remove the critics' objections to overturning popular beliefs, he playfully banned readers who never asked themselves about their surroundings. "Vulgar and ignorant men [find] these things [to be] easily understandable, because according to them, feeling sensations repeatedly and understanding them are the same thing," he observed, "but I do not intend to speak with such persons." That gave him a good excuse to go into the Galilean and anti-Aristotelian explanation of hot and dry, to which he added a corpuscular view supported with long and amusing citations from his teacher Marchetti's translation of Lucretius.

Montanari went even further. In his Galilean dialogue on *The Power of Aeolus*, concerning a tornado that had "viciously struck many towns and villages in the territories of Mantua, Padua and Verona," he tried to provide an easily accessible extended discussion that arrived, by cumulative conclusions, not at quantitative physics, but at the Cartesian theory of vortices. He even tried to surpass Galileo by using repeated references to the main topic to ensure that the reader would not get lost in his digressions on such other commonplace phenomena as the diffusion of water in waterfalls and the sound of a cracking whip—all of which could be explained by Cartesian corpuscularism. And to make sure the reader got his general point about the place of virtuosi in providing explanations for everyday phenomena, he included himself not as an unnamed academic, which had been Galileo's technique in the *Dialogue Concerning the Two Great World Systems*, but as one of the named interlocutors.[81]

At first, journals contributed mainly by establishing the qualifications of the practitioners who took up their new roles as providers of explanations. Thus, when the Roman *Giornale de' letterati* assured the reader that Ippolito Magnani was "a surgeon no less diligent in the cure of illnesses than sagacious and curious in the investigation of their causes,"[82] this was no mere concession to Baroque rhetoric practice. It was an acknowledgement that if science was to be dedicated to "public utility," some knowledge about who provided it seemed important.[83] The same went for the journal's reference to Salvatore Serra as "exceedingly expert in optics and other sciences."[84]

The independent journal article emerged in 1696 when Albrizzi began to include contributions whole and uncut. Practitioners could now publicize the main ideas of contemporary science and show themselves as the custodians of knowledge. Fardella took the cue to provide an exposé of one of the hottest topics of the day—Cartesian philosophy and its relation to Catholic orthodoxy. "The novelty of Descartes can be summed up," he explained, "in a . . . new method of philosophizing . . . the mixing of physical and geometrical matters . . . that leads us spontaneously to the knowledge of many basic truths." The worst danger was not in the philosophy's opposition to Christianity but in not taking the philosophy seriously enough. "Many have said that it seems adaptable," he complained, "to the weak and certain capacity of any uncultivated little girl."[85] He then explained how science could be an essential aid to faith by leading the mind where theology could not, and he ridiculed virtuosi who tried to mix the two. "What kind of philosophizing is this?" he said of one author. "First guided by Descartes he asserts that he understands with clarity and full evidence the existence of God and of his attributes. Then he abandons science and relies on faith alone by saying he firmly believes this."[86]

Finally, practitioners took up the journals in an appeal to the public for support in their scientific debates. Albrizzi's open-door policy allowed them to use every tool available for convincing a popular audience about the validity of their conclusions—even if this meant neglecting legitimate scientific evidence in favor of rhetorical slight-of-hand. Vallisneri was exemplary. Sometimes he did not even bother to include the place of publication in reviews of works he opposed, since he had no interest in encouraging others to read them. Other times he reviewed works over a decade old just so he could bring up his favorite theses. "This book has traveled over from France," he began one such review, "just to damage the good credit of medicine in Italy."[87] In reviews of books he favored because they reflected the views of the moderns, he added to the impression of cultural depth and informativeness by squeezing as many names as he could into one sentence. "[The author] invites all physicians to defend their opinions," he noted of one, "before Apollo himself, Bacon, and Boyle, elected as his assessors, and among his Councilors, Galileo, Redi and Gassendi."[88] In reviews of his own books, he added rhetorical force to the original. In one, he replaced the documentary and empirical evidence of the original with assurances that such-and-such "proves that" or "demonstrates" and conclusively "refutes Aristotle" on spontaneous generation, avoiding the reader's personal evaluation of his evidence and so remaining at all times in full control of the significance of his presentation. Likewise, physician Lodovico Testi insisted in a review that his own book contained sufficient "physical reasons" to "prove that the air of Venice is entirely healthy," and reserved his substantive exposé of the book's arguments to a letter responding decisively to the objections of a critic.[89] Who could doubt his word?

Finally, practitioners took over the task of writing textbooks, often experimenting with new and more appealing modes of presentation. Alessandro Pascoli, professor at the Sapienza in Rome, published a "new method" intended to introduce his readers to the latest acquisition of the search of the Italian practitioners for a philosophical basis for empirical research: introspection. In the first part, "an essay on metaphysics in the style of Descartes," he showed that textbook writing could mean something more than just cribbing from the most authoritative or the most fashionable authors in the hands of someone who happened to be a specialist in the field. And his approach was so promising that his printer saluted the "universal pleasure that the works of Signor Pascoli encounter" and dared to print them "at my own expense."[90]

The late seventeenth-century system of scientific communications permitted practitioners in the fields of natural knowledge to reconcile themselves to their new institutional situations. They could not expect instant recognition from their institutions now that science was divided between the universities, the Church, and the various governments, none of which was exclusively devoted to science. But the emergence of a market for modern natural knowledge for the moment signified that no one was going to bother them any more about what they wrote—not, at least, for the reasons that had affected Galileo and his followers. They could ensure recognition among their fellows in very different institutional situations abroad by numerous outlets, from the epistolary networks to the journals to the innumerable independent forms, any of which could be combined with the others for innumerable repeat publications. They could depend on the journals to ensure that none of this activity would go unnoticed. They could procure social recognition among fellow-subjects of a state by the increasingly clear definition of themselves and their activities that they projected in their publications aimed at a popular audience. And with recognition at home and abroad, rewards from their institutions would surely follow.

By the end of the century, the process was almost complete, and Bianchini's assessment of the career of Montanari was emblematic of the changes that had taken place. "His profession ... was not to sit in the shadow of a [university] chair practically at leisure, but to make the speculations of his mind useful for life and direct all of the theorems of science to this problem: to make science the minister of public felicity rather than a testimony of private effort." None of the late seventeenth-century practitioners could have disagreed.[91]

In spite of what modern scholarship might regard as journalism's superiority as a communications tool, it never completely took over. Until well into the eighteenth century, natural science practitioners in Italy never sought forms of journalism more similar to the *Philosophical Transactions of the Royal Society*, which, by the admission of Sir Robert Moray, remained intentionally distinct from continental journalism by its inclusion of more reports on observations and experiments.[92] The peculiar exigencies of the Italian market demanded that accomplishments in the natural sciences should be set alongside accomplishments in every other field of learning. Journalism was assimilated into, instead of supplanting, an existing system, and within this system it added further motivations to those that were coming from educational institutions, as we shall see.

Notes

1. *Correspondence of Marcello Malpighi*, ed. Howard B. Adelmann, 5 vols. (Ithaca: Cornell University Press, 1975), 1: 115, 19 January 1661. Concerning Malpighi, still fundamental is Howard B. Adelmann, *Marcello Malpighi and the Evolution of Embryology*, 5 vols. (Ithaca: Cornell University Press, 1966). See also Domenico Bertoloni Meli, ed., *Marcello Malpighi, Anatomist and Physician* (Florence: Olschki, 1997).

2. *Lettere inedite di huomini illustri*, ed. Angelo Fabroni, 2 vols. (Florence: F. Moucke, 1773-5), 2: 161: "Impari a spese di Galileo, che patì tante contraddizioni, e molte per averla presa con questo e quello."

3. Gregorio Leti, *Italia Regnante*, 4 vols. ("Valenza" [=Venice]: Vincenzo Guerini, 1675-6), 2: iiir.

4. *Giornale de' letterati* [hereafter, *GLR*] (Rome: 1674), 97-98. Translation is from my *Baroque Italy: Selected Readings* (New York: Garland, 1995), 582. My view of the situation in France is in part based on A. Beaulieu, "Les réactions des savants français au début du dix-septième siècle devant l'heliocentrisme de Galilée," in Paolo Galluzzi, ed., *Novità celesti e crisi del sapere. Atti del convegno internazionale di studi galileiani, 1982* (Florence: Giunti Barbera, 1984), 373-82.

5. Porzio's Cartesianism is analyzed in Alessandro Dini, *Filosofia della natura, medicina, religione: Lucantonio Porzio, 1639-1724* (Milan: Angeli, 1985), 14-24. The "atheist" trials are the subject of Luciano Osbat, *L'Inquisizione a Napoli. Il processo agli ateisti, 1688-1697* (Rome: Edizioni di storia e letteratura, 1974). The Malpighi incident is reported in *Correspondence of Marcello Malpighi*, 1:400.

6. *Lettere inedite*, 1: 117, to Prince Leopoldo, 11 November 1658.

7. *Correspondence of Marcello Malpighi*, 1: 400, 14 February 1669.

8. A. Rupert Hall, *Philosophers at war: the quarrel between Newton and Leibniz* (Cambridge, 1980), chap. 3. Mail services are outlined in Bruno Caizzi, *Dalla posta dei*

re alla posta di tutti. Territorio e comunicazioni in Italia dal XVI secolo all'Unità (Prato: Istituto di studi storici postali, 1993).

9. Research problems with the Magliabechi correspondence are outlined in Manuela Doni Garfagnini, *Lettere e carte Magliabechi: Regesto*, vol. 1 (Rome: Istituto storico italiano per l'eta moderna e contemporanea, 1981). The best *mis à point* of Magliabechi's role is Françoise Waquet, "Antonio Magliabechi: nouvelles interpretations, nouveaux problèmes," *Nouvelles de la république des lettres* 1 (1982): 173-88. In addition, *Correspondence of Marcello Malpighi*, 4: 1664 (quote). In addition, here, Paul Dibon, "Les échanges épistolaires dans l'Europe savante du dix-septième siècle," *Revue de synthèse* 97 (1976), 47.

10. Albert Van Helden, "The Accademia del Cimento and Saturn's Ring," *Physis* 15 (1973): 237-59.

11. *Correspondence of Marcello Malpighi*, 4: 1579, 25 February 1690.

12. Jean Huguetan, *Voyage d'Italie curieux et nouveau* (Lyons: T. Amaubry, 1681), 318, 283, 248. Galileo's personal campaign in Rome is elucidated by Richard S. Westfall, "Science and Patronage: Galileo and the Telescope," *Isis* 76 (1985): 11-30.

13. *Correspondence of Marcello Malpighi*, 1: 110, 29 December 1661.

14. *Correspondence of Marcello Malpighi*, 1: 115, 19 January 1661.

15. Stefano Degli Angeli, *Considerazioni sopra la forza di alcune ragioni fisiomatematici addottate dal M.R.P. Giovanni Battista Riccioli della Compagnia di Gessù nel suo Almagest Nuova e Astronomiae Riformata contro il sistema copernicano* (Venice: 1667). More details about Degli Angeli are in L. Tenca, "Stefano Degli Angeli," *Atti dell'Accademia delle Scienze dell'Istituto di Bologna*, ser. 2, vol. 5 (1958): 194-207.

16. Letter of 11 October 1670, quoted in Mario Saccenti, *Lucrezio in Toscana* (Florence: Olschki, 1966), 72. My analysis diverges from Paolo Galluzzi, "Libertà scientifica, educazione e ragion di stato in una polemica universitaria pisana del 1670," in *Atti del XXIV congresso nazionale di filosofia*, L'Aquila, 28 aprile-2 maggio 1973 (Rome: Società filosofica italiana, 1973), vol. 2 pt. 2.

17. Saccenti, *Lucrezio in Toscana*, 103.

18. The best account of this, and the material in the next two paragraphs, is Jean Dietz Moss, *Novelties in the Heavens. Rhetoric and science in the Copernican controversy* (Chicago: University of Chicago Press, 1993). Galileo's comment is in *Edizione nazionale delle opere di Galileo Galilei*, 3rd edition, 20 vols. (Florence: Giunti Barbera, 1967), 9: 334; those of his opponents are in *Ibid.*, 15: 56.

19. Contemporary rhetorical decorum in science was the object of Sforza Pallavicino, *Trattato dello stile e del dialogo, ove nel cercarsi l'idea dello scrivere insegnativo, discorresi partitamente de' veri pregi dello stile sì latino come Italiano* (Rome: Mascardi, 1662 [1st impression: 1661]), chap. 30.

20. Saccenti, *Lucrezio in Toscana*, 89.

21. For reasons I explain in the next chapter.

22. *Correspondence of Marcello Malpighi*, 1: 38, 31 July 1660.

23. *Correspondence of Marcello Malpighi*, 4: 1553, 17 November 1689. Concerning Bellini, there is Giorgio Weber, *L'anatomia patologica di Lorenzo Bellini anatomico (1843-1704)* (Florence: Olschki, 1998).

24. I differ here from the conclusions of Maurizio Torrini, "Due galileiani a Roma: Raffaele Magiotti e Antonio Nardi," in *La scuola galileiana: prospettive di ricerca, atti del Convegno di S. M. Ligure, 1978,* ed. Gino Arrighi (Florence: La Nuova Italia, 1979), 53-63.

25. *Correspondance de Marin Mersenne,* ed. Paul Tannery and Cornelius De Waard, 17 vols. (Paris: Éditions du C.N.R.S., 1969), 6: 42, March 1636; 235, 20 June 1637; 258, 17 May 1637, and Descartes, *Oeuvres*, ed. Charles Adam and Paul Tannery, 2nd ed. (11 vols., Paris: Vrin, 1971-), 6: 515-18.

26. Francesco Eschinardi, *Microcosmi physico-mathematici, seu compendii, in quo clare et breviter tractantur praecipuae Mundi partes, Coelum, Aer, Aqua, Terra: eorum praecipua accidentia*, 1 (the only one published), (Perugia: 1658), quoted here at the unpaginated introduction. Biographical details are in William E. Knowles Middleton, "Science in Rome: 1675-1700, and the Accademia Fisico-Matematica of Giovanni Giusto Ciampini," *British Journal for the History of Science* 8 (1975): 138-54; and in the entry by M. Mucillo in *DBI* 43 (1993): 273-4.

27. Claude Bérigard, *Circulus pisanus. De veteri et peripatetici Philosophia, in Aristotelis libros octo Physicorum, quattuor de coelo, Duos de Ortu et interritu, Quattuour de meteoris et Tres de anima* . . . (Padua: Frambotti, 1661), 18. Bérigard is principally known for his opposition to Galileo's cosmography, which is the basis for recent studies on him by Maria Laura Soppelsa, *Genesi del metodo galileiano e tramonto dell'Aristotelismo nella Scuola di Padova* (Padua: Antenore 1974), 92-112; M. Bellucci, "La filosofia naturale di Claudio Berigardo," *Rivista critica di storia della filosofia* 26 (1971): 363-411; and Giorgio Stabile, "Il primo oppositore del *Dialogo*: Claude Bérigard," *Novità celesti*, 277-82, and his *Claudio Berigard, 1592-1663. Contributo alla storia dell'atomismo seicentesco* (Rome: Istituto di filosofia dell'Università, 1975).

28. Paolo Galluzzi, "Lettere di Giovanni Alfonso Borelli ad Antonio Magliabechi," *Physis*12 (1970): 277.

29. Giovanni Serpetro, *Mercato delle meraviglie* (Venice: per il Tomasini, 1653), unpaginated introduction. See Corrado Dollo, *Modelli scientifici e filosofia nella Sicilia spagnola* (Naples: Guida, 1984), 132-36. The earlier tradition of collections of secrets is the subject of William Eamon, *Science and the Secrets of Nature* (Princeton: Princeton University Press, 1994).

30. The work of the Jesuit investigators—not to be confused with the textbook compilers—is examined by Paolo Galluzzi, "Galileo contro Copernico: il dibattito sulla prova 'galileiana' di Giovanni Battista Riccioli contro il moto della Terra," *Annali del Istituto e Museo di Storia della Scienza, Firenze* 2, no. 2 (1977): 87-97; Maurizio Torrini, "Giuseppe Ferroni Gesuita e galileiano," *Physis* 15 (1973): 411-23.

31. Modern textbooks are analyzed by Roberto Maiocchi, "Il segreto di Pulcinella: la vittoria dell'atomismo attraverso la manualistica fisica," *Società e storia* 10 (1987): 17-52; 301-32. Earlier ones are analyzed by Patricia Reif, "The Textbook Tradition in Natural Philosophy, 1600-1650," *Journal of the History of Ideas* 30 (1969): 29. Jesuit teaching in Italy in this period is analyzed by Gabriele Baroncini, "L'insegnamento della filosofia naturale nei collegi italiani dei Gesuiti, 1610-1670: un esempio di nuovo aristotelismo," in Gian Paolo Brizzi, ed., *La "Ratio Studiorum": modelli culturali e pratiche educative dei Gesuiti in Italia tra Cinque e Seicento* (Rome: Bulzoni, 1981), 163-216; Romano Gatto, *Tra scienza e immaginazione. Le matematiche presso il collegio gesuitico napoletano, 1552-1670* (Florence: Olschki, 1994); and Ugo Baldini, *"Legem impone subactis," studi su filosofia e scienza dei Gesuiti in Italia, 1540-1632* (Rome: Bulzoni, 1992), 401-69, covering the period from 1600 to 1660.

32. Giattini, *Physica* (Rome, 1653), 465. There is no study on Giattini.

33. Giattini, *Physica*, 621.

34. Giattini, *Physica*, 630-31.

35. Giattini, *Physica*, 615.

36. *Lettere inedite*, 2: 249, 5 September 1674, to Prince Leopold.

37. *Correspondence of Marcello Malpighi*, 4: 1478, 13 July 1689, to Scipione Borghese.

38. On Boccone's laments, C. Dollo, *Modelli scientifici e filosofia nella Sicilia spagnola*, 69. Rossetti's are in *Lettere inedite*, 2: 249, 5 September 1674, to Pietro Leopoldo.

39. The seventeenth-century industry is still relatively unstudied, although there are some rudimentary indications in Alfonso Mirto, *Stampatori, editori, librai nella seconda*

metà del Seicento (Florence: Centro editoriale toscano, 1984).

40. The incident is recorded in *Correspondence of Marcello Malpighi*, 3: 1276.

41. Jean-Michel Gardair, *Le "Giornale de' letterati"de Rome (1668-1681)* (Florence: Olschki, 1984). Also helpful on this and the other Italian journals is Giuseppe Ricuperati, "Giornali e società nell'Italia dell'Ancien Régime," *La stampa italiana dal Cinquecento al Settecento*, V. Castronovo and Nicola Tranfaglia, eds. (Bari: Laterza, 1976), 67-372; May Katzen, "The Changing Appearance of Research Journals in Science and Technology: an Analysis and a Case Study," in A. J. Meadows, ed., *The Development of Scientific Publishing in Europe* (New York-Amsterdam: Elsevier Science Publishers, 1980), 177-236.

42. *GLR* (1674): 151.

43. *GLR* (1669): 127.

44. *GLR* (1672): 9.

45. *Correspondence of Marcello Malpighi*, 4: 1600, 14 May 1690; 1434, 30 January 1688.

46. *Correspondence of Marcello Malpighi*, 3: 1088, 22 October 1685.

47. Baglivi himself sent his *Medicina, pars altera* (Avignon, 1687) for "onorata menzione." *The Baglivi correspondence from the library of Sir William Osler*, ed. Dorothy M. Schullian (Ithaca: Cornell University Press, 1974), 102.

48. Montanari, *La livella diottrica* (Venice, 1680), dedication.

49. Montanari, *Specolazioni fisiche sopra gli effetti di quei vetri temperati che rotti in una parte si risolvono tutti in polvere, esposti in due lettere* (Bologna, 1671), preface by Agostino Fabri, a student of Montanari, "Al Discreto Lettore," n.p.

50. Cassini's broadsheets are *Martis circa axem proprium revolubilis observationes Bononiae* (Bologna, 1666); *De periodo quotidianae revolutiones Martis* (Bologna, 1666); *De aliis Romanis observationibus macularum Martis* (Bologna, 1666). Serra's reply was in *Martis Revolubilis observationes romanae ab affictis erroribus vindicatae* (Rome, 1666). All are in Venice, Biblioteca Nazionale Marciana, Misc. 871. A later example of the same genre was Giovanni Francesco Vanni's *Specimen liber de momentis gravium* (Rome, 1684).

51. Paolo Boccone, *Museo di fisica e di esperienze, variato e decorato di osservazioni naturali e note medicinali, e ragionamenti, secondo i principi de' moderni* (Venice, 1697), 39. The last quote is from p. 21.

52. *Saggi di naturali esperienze* (Florence: Cocchini, 1667). In this regard, Paolo Galluzzi, "L'Accademia del Cimento: 'gusti' del Principe, filosofia e ideologia dell'esperimento," *Quaderni storici* 16 (1981): 788-844, although I disagree with the conclusion.

53. Geminiano Montanari, *Pensieri fisico-matematici sopra alcune esperienze fatte in Bologna nell'Accademia Filosofica eretta dall'Illustrissimo e Reverendissimo Signor Ab. Carlo Antonio Sampieri intorno diversi effetti di liquidi in canuccie di vetro e altri vasi* (Bologna: Manolessi, 1667). Another similar work was Domenico Guglielmini, *Riflessioni filosofiche dedotte dalla figura dei sali* (Bologna: Pisarri, 1688).

54. A few biographical details on Albrizzi are from the entry by Giorgio E. Ferrari in *Dizionario biografico degli italiani* 2 (1960): 58-59; and from M. Lanaro, "Accademie ed editoria: l'attività degli Albrizzi a Venezia," in *Accademie e cultura: aspetti storici tra Sei e Settecento* (Florence: Olschki, 1979), 227-72.

55. [Giovanni Domenico Cassini and Geminiano Montanari], *Relazione dell'esperimenze fatte intorno alla trasfusione del sangue* (Rome, 1668), reprinted in Bologna the same year.

56. On which, I refer to David A. Kronick, *A History of Scientific and Technical Periodicals: The Origins and Development of the Scientific and Technical Press, 1665-1790* (Metuchen, N.J.: Scarecrow Press, 1976); and William D. Garvey, *Communication: The*

Essence of Science (New York: Pergamon Press, 1979), 5.

57. See Susana Gómez López, *Le passioni degli atomi. Montanari e Rossetti: una polemica tra galileiani* (Florence: Olschki, 1996); Massimo Bucciantini and Maurizio Torrini, eds., *Geometria e atomismo nella scuola galileiana* (Florence: Olschki, 1992).

58. *Correspondence of Marcello Malpighi*, ii, 155, 30 March 1663.

59. Paracelsianism is analyzed in the articles by Marco Ferrari and Paolo Galluzzi in *Scienze, credenze occulte, livelli di cultura, convegno internazionale di studi, Firenze, 26-30 June, 1980* (Florence: Olschki, 1982), 21-30, 31-62.

60. The quote from Malpighi is in *Correspondence of Marcello Malpighi*,1: 70. In addition, Maurizio Torrini, "Uno scritto sconosciuto di Lionardo Di Capua in difesa dell'arte chimica," *Bollettino del Centro di Studi Vichiani* 4 (1974): 126-39; Eugenio Garin, *Dal Rinascimento all'Illuminismo* (Pisa: Nistri-Lischi, 1970), 135-44; Walter Bernardi, *Le metafisiche dell'embrione: scienze della vita e filosofia da Malpighi a Spallanzani, 1672-1793* (Florence: Olschki, 1986), 68-70, 112-19.

61. The purely epistemological aspects of the shift are traced in Cesare Vasoli, *L'enciclopedismo del Seicento* (Naples: Bibliopolis, 1978). Bacchini's Weltanschauung is the subject of Arnaldo Momigliano, "Mabillon's Italian Disciples," *Essays in Ancient and Modern Historiography* (Middletown, Conn.: Wesleyan University Press, 1977), 277-93.

62. *Galleria di Minerva* [hereafter, *GM*], (1696): 368. A useful study is Donatella Lauria, *Agostinismo e cartesianismo in Michelangelo Fardella* (Catania: N. Giannotta, 1974).

63. *Giornale de' letterati* (Parma/Modena: 1697): 31.

64. This is the main argument of Bernardi's *Le metafisiche dell'embrione*. In addition, Lionardo Di Capua: *Parere divisato in otto ragionamenti, ne' quali particolarmente narrandosi l'origine e il progresso della medicina, chiaramente l'incertezza della stessa si manifesta* (Naples: Bulifon, 1681), 471. Vallisneri: "Dialogo sopra la curiosa origine di molti insetti," 298. Capucci and Cestoni: *Correspondence of Marcello Malpighi*, 1: 411; 5: 1120.

65. Maurizio Torrini, *Tommaso Cornelio e la ricostruzione della scienza* (Naples: Guida, 1977).

66. On the role of this journal, see my *Science, Politics and Society in Eighteenth-Century Italy* (New York: Garland, 1991).

67. The exchange between Waller and Malpighi is in *Correspondence of Marcello Malpighi*, 5: 1916-18. Bellini's difficulties in satisfying his correspondent are recorded in *Ibid.*, 3: 1276.

68. "Due osservazioni dell'eclisse lunare . . . la prima fu fatta in Bologna dal Sig. Conte Herede Zani e Pietro Mengoli e . . . l'altra . . . [in] Ginestreto, castello di Pesaro . . . dal Sig. Ab. Giovanni Francesco Laurenzi," *GLR* (1674): 120, with a table comparing them.

69. Marie Boas Hall, "The Royal Society's role in the diffusion of information in the seventeenth century," *Notes and Records of the Royal Society of London* 29 (1975): 188.

70. *GLR* (1669): 104.

71. *Genio dei letterati appagato* (1706), n.p.; *GM* (1696), "Ai lettori," n.p.

72. Some examples from ASV, *Petizion: Inventari*, filza 386.51.73 n.d., but 1680s; 385.51.82, 26 February 1685; 390.55.32, 1689; 386.51.39, 23 March 1684; 196.61.29, 5 April 1699.

73. ASV, *Petizion: Inventari*, filza 356.21.74, 5 August 1639.

74. ASV, *Petizion: Inventari*, filza 366.30.90, 31 November 1657.

75. ASV, *Petizion: Inventari*, filza 388.53.29bis, 17 January 1688. Chechel's large collection of paintings is listed in Simona Savini Branca, *Il collezionismo veneziano nel Seicento* (Florence: Olschki, 1965), 141.

76. ASV, *Petizion: Inventari*, filza 339.64.5, 8 April 1702.
77. ASV, *Petizion: Inventari*, filza 364.28.81, 30 July 1653.
78. *GM*, (1697), 237 and 383. The quote is from *Ibid.*, 301: "Come se vorrai dimostrare che l'oro è nulla. Di zero fia zero fa nulla: la r è lettera canina, perchè la pronunziano i cani; quando per un osso fremono fra di loro. Or appunto per un nulla contrastano tutti gli uomini."
79. Later collected in the *Opuscoli vari di Francesco Redi* (Florence: Piero Matini, n.d., but probably 1674).
80. Giuseppe Del Papa, *Della natura del umido e del secco* (Florence: Vincenzo Vangelisti, 1681), 7.
81. Geminiano Montanari, *Le forze dell'Eolo: dialogo fisico-matematico sopra gli effetti del vortice, o sia turbine, detto negli stati Veneti la "bisciabuova" che il giorno 29 1686 ha scosso e flagellato molte ville e luoghi nei territori di Mantova, Padova, Verona, eccetera, opera postuma* (Parma: A. Poletti, 1694). Other aspects of this work are noted in Salvatore Rotta, "Scienza e 'pubblica felicità' in Geminiano Montanari," in *Miscellanea Seicento* (Florence: Le Monnier, 1971), 2: 64-204, on which I rely for other information regarding Montanari.
82. *GLR* (1668): 139.
83. For example, *GLR* (1674): 57: "Il buon genio del Sig. Cardinal de' Medici [i.e., Prince Leopoldo] a favorire e promuovere gli studi e matematiche e filosofiche ed altri fa che tra le molte gravissime occupazioni sue trovi anche tempo di pensare a quel che può servire ai virtuosi per privata e pubblica utilità."
84. *GLR* (1668): 79.
85. *GM*, (1697): 42.
86. *GM*, (1697): 203.
87. *GM*, (1696): 387, in reference to [Jean Devaux] *Le médecin de soi-même* (Leyden: 1682). In addition, *GM*, (1700): 98.
88. *GM*, (1706): 322.
89. *GM*, (1697): 282, 290.
90. Alessandro Pascoli, *Nuovo metodo per introdursi ad imitazione dei geometri con ordine, chiarezza e brevità nelle più sottili questioni di filosofie, metafisiche, logiche, morali, e fisiche* (Venice: Andrea Poletti, 1702), printer's preface.
91. Francesco Bianchini, Introduction to Montanari, *Le forze dell'Eolo*, n.p.
92. Christiaan Huygens, *Oeuvres Complètes*, 22 vols. (The Hague: M. Nijhoff, 1888-1950), 5: 234. On this, Wolfgang Van den Daele, "The Social Construction of Science. Institutionalization and Definition of Positive Science in the Latter Half of the Seventeenth Century," in Everett Mendelsohn, Peter Weingart, and Richard Whitley, eds., *The Social Production of Scientific Knowledge, Sociology of the Sciences: a Yearbook* 1 (Dordrecht-Boston: D. Reidel, 1977), 31.

Chapter 4:
Nature and the Universities

Not all students attended early modern universities with the intention of broadening their intellectual horizons or engaging in free inquiry into the deepest currents of contemporary knowledge. Some of them simply wanted a degree. As Federico Cesi noted, "Most scholars follow the professions more apt for this approach, that is, law and medicine, the latter for the daily fees collected from house to house, and the former for similar fruit from employments, honors, ministries with princes, attorneyships and procuratorships."[1] Nor were administrators inclined to discourage students from such aspirations—especially as populating the growing state apparatus with reliable functionaries and providing skilled physicians seemed like a reasonable purpose for the institutions that existed in their states. The demand for instruction, formed of students' ambitions and administrators' desires, Cesi believed, powerfully conditioned the supply of scientific ideas. Intellectual development, he suggested, would be best served if real seekers after natural knowledge were to withdraw from all worldly concerns and compete only for each other's esteem. As patron and founder of the early seventeenth-century Roman Accademia dei Lincei, an exclusive body uniquely dedicated to the production of ideas, he set an agenda that would be followed with variants by founders of scientific academies throughout Europe.

Modern historians have focused more on the role of the state than on that of the students in creating the particular marketplace in which professors exercised their duties. Administrators, inspired by fantasies about a fully-disciplined society and infused with counter Reformation notions about the advantages of cultural hegemony, this scholarship says, welcomed opportunities to influence the university curricula. And where they could, they ensured that these did not depart from the deeply rutted tracks of medieval scholasticism.[2] Over the long term, no one could wonder that the universities foundered in "the deepest decadence," only to rise again when enlightened bureaucrats revived them in the late eighteenth century. If universities in northern Europe only maintained a modest edge over their cisalpine counterparts across the period, the standard view continues, their defects were compensated by the presence of more solid academic organizations than the short-lived Lincei, the transitory Accademia del Cimento in Florence, and the embattled Accademia degli Investiganti in Naples.[3]

And if the scientific role of universities in Italy were to be judged exclusively based on the professors' announced teaching programs, there would be no cause to doubt the standard view.[4] Professors in the medical faculties, where most material of a scientific nature was taught, gave a three-year course comprising Hippocrates' *Aphorisms,* Galen's *Methodus medendi,* and various treatises from Avicenna's *Canon.* Philosophy professors taught a three-year course combining elements from Aristotle's *De Anima* I, II, and III, *Physics* I, II, and VIII, *On Generation and Corruption,* and *De Coelo.* As late as the first decade of the eighteenth century, a typical examining board in medicine and philosophy called for an explanation of *Aphorism* 8 of Hippocrates and *De Anima* "lib. 2 tex. 67" (i.e., 418b2) of Aristotle.[5] A better corroboration could scarcely be imagined for Cesi's observation that professors sought only to expound "the more common views of the ruling sect" of philosophers.

However, recent work has shown that the statutes are not a reliable guide to what went on in early modern classrooms.[6] How could they be? Even supposing that lessons were intended to be commentaries on an ancient text, as they actually were in many cases, not all commentaries were alike.[7] Some were anything but flaccid compilations of consensus-inducing ideas. Pietro Pomponazzi taught his early sixteenth-century students at Padua what he believed to be the real meaning of Aristotle's metaphysics and *De anima*—i.e., the mortality of the soul and the finiteness of divine power. Bernardino Tomitano devoted his Padua lessons on logic to new methods of reasoning adaptable to the arguments used by mid-sixteenth-century humanists and natural philosophers. Giambattista Da Monte audaciously asserted, in his medical teaching of the same period, that the Galenic notion of *temperatura* or *complexio*—that is, the balance of the four qualities of hot, cold, wet, and dry—was identical with substantial form and thus with the soul, taking the opportunity to expound the latest corpuscular theories of matter. And he spoke for all the rest when he explained why it was worthwhile to adhere to the traditional form of the commentary on an ancient author—in this case, Avicenna: "Someone may say . . . 'why are you expounding a book that you attack so vigorously?'" he noted. "In the first place, I am not attacking the whole book, but only those things in it that seem to me reprehensible; besides, I am not doing anything new. For Galen never wrote better commentaries than on books that he opposed."[8]

And what has been discovered recently concerning classroom teaching gives evidence for a far different picture of the role of universities from the one suggested by the critics at the time and by the last generation of historians. In fact, many of the great natural philosophers in Italy were closely involved with universities. Galileo, it is said, drew inspiration from "the continuing vitality of the universities in [his] lifetime."[9] His followers, moreover, including Benedetto Castelli, Giovanni Alfonso Borelli, and Marcello Malpighi, trained sturdy groups of disciples from their positions in professorships at Padua, Bologna, Pisa, and Messina.[10] All of these would have shared the satisfaction of Lorenzo Bellini, medical professor at Pisa, at "seeing the university each year more filled with my dependents, without my even saying anything, and now the principal chairs of practical and theoretical medicine are filled, with universal approval, by my scholars; likewise also the chairs of Simples, Logic, and Philosophy."[11] They too could count on the attainment of such a level of philosophical beatitude that they "could not find [themselves] in any kind of circumstances, however strange, in which [they] should not find a huge wealth of things that could keep [them] always supplied with new, unusual, extraordinary, and immense knowledge, of uncountable value." And they shared with Bellini a social position in which "no one can detract from our well-being, our fame, and our noble satisfactions."

Important work continued to be done in the wake of Galileo's condemnation, recent studies have shown, at least in part because university intellectual activity never really slowed down.[12] What the following discussion will try to show is that the influence of a marketplace upon professors' activities did not produce uniformly negative results either in education or in science. Instead, a varied demand for instruction and scientific expertise was met by an equally varied supply.

Whether the professors followed it scrupulously or not, the statutory curriculum continued to reflect a significant trend of ideas about the application of natural knowledge to public health in the early modern period. All over Italy, Galen and Hippocrates remained authoritative in medicine mainly because no one could agree on anything else. The possible alternatives, from Paracelsus to van Helmont to Giorgio Baglivi, were impugned by many on theoretical as well as metaphysical grounds. Indeed, the universities themselves, as we saw at the end of the last chapter, were the battlefields upon which vitalists, mechanists, and those belonging to the various intermediate strains fought for intellectual mastery. The story told by Malpighi about the solemn burning of his house by those who disagreed with his microbiological observations may have been his own invention. But Morgagni's move to the University of Padua to escape his "Galenist" adversaries in Bologna was no myth.[13] The professors themselves occasionally resorted to miracle drugs in spite of statutory fulminations against these practices—as in the case of Alessandro Knips Maccope at Padua, who earned such a reputation by his use of mercury in the early eighteenth century that patients came to his studio from all over Europe "as to the altar of Æsculapius."[14] Since the best way to cure disease was often better arrived at by forgetting the theory for a moment and using pure good sense and intuition, most professors—"ancient" and "modern" alike—agreed that Galen and Hippocrates were at least not a useless foundation upon which to build solid medical knowledge. Thus, the traditional curriculum continued to satisfy the needs of an important segment of the university population throughout our period. Indeed, anyone who wished to enter the medical profession had to take an exam; and anyone who took the exam could be certain of being asked about the standard materials.

This is not to say there was no possibility of innovation. When most of the curricula were codified in the fifteenth century, the possibility of considerable improvements within each branch of study was fully taken into account. Professors had to decide, for example, which to follow when two such authoritative authors as Avicenna and Galen differed on a particular point. Whenever the authorities were obscure, they had to resort to their own ingenuity to solve the difficulty. They could even call upon the guidance of some of the more recent treatises in a particular field—in physics, for example, such thirteenth-century treatises as Giovanni Sacrobosco's *Sphaerae* or, in medicine, Bruno da Longoburgo's *Chirurgia minor*.[15] Eventually, printed versions of the standard commentaries on texts relieved the professors of the need to provide students with all the possible material they might need about a text and permitted them to experiment instead with their own views.[16] However, in spite of these latitudes, most of them took their major responsibility to be that of infusing students with the elements of the ancient medical and legal learning in much the same way as the professor of humanities infused them with the elements of ancient rhetorical techniques. At Bologna, a "preliminary examination" of the extant lessons shows that they constitute "an Aristotelian *continuum*."[17]

Nor were these requirements likely to change significantly as long as the local colleges of physicians responsible for examining candidates for entry continued to insist upon them.[18] The examining boards in university towns were made up of the local professional colleges' most prominent members, who were just as concerned to defend their ancient privileges as were their peers in other non-university cities. They valued their unique role in examining the professional competence that would qualify candidates for membership both in their own and in any of the colleges in non-university towns. The knowledge whose acquisition they wished to judge was the same that they had acquired as students in the local university.[19] They insisted upon performing the examination with a minimum of assistance from the separate divisions of their college or the entirely separate colleges devoted to the university professors.[20]

The evolving state bureaucracies took care to defend the power and prestige of local professional corporations. In Tuscany, the government generally avoided interfering in the affairs of either the professors or the physicians. Meanwhile, it did everything it could to add its own authority to that of the statutes in protecting corporate privileges. To the corporations' concerns about the qualifications of candidates and the procedures for holding disputations, it responded by appropriate legislation.[21] Wherever there were universities, governments emanated decree after decree warning professors not to fill students' heads with useless ideas and enjoining them to teach the standard texts recommended by the college statutes and employed by the examining boards.[22]

However, in defending the corporations, the Italian governments rarely lost sight of patronage goals that might increase their own power and prestige. Universities, they found, were fit objects for the initiatives they pursued in other programs connected with the arts and sciences. Cosimo I of Tuscany was by no means the only ruler to exclaim that "the time of peace" was ideal for "turning to the cause of university reform."[23] Nor was he alone in boasting to another ruler about the virtues of the nearby university, in this case, Pisa. The late fifteenth- and early sixteenth-century Italian wars delayed reform in most places. But the subsequent establishment of the first relatively stable Italian state system provided those governments that had remained and those new governments that were then established with the opportunity to take up once again the question of reform. New university buildings, the overhaul of statutes, and the creation of special university administrative bodies were equated with good politics everywhere.

The Venetian government was exemplary. Almost as soon as it had regained Padua in 1517 along with many of the other territories it had lost at the battle of Agnadello, it took up the question of university reform with the same energy that it applied to all other Terraferma operations. It centralized university administration—which had previously been diffused between the Senate, the Venetian governor in Padua and numerous separate student and faculty representatives—into a single magistracy, the three Riformatori dello Studio. The Riformatori in turn consolidated all university activities into one magnificent location by converting the Palazzo del Bò into a classroom building. They coordinated the relationships between the various parts of the university by ordering an expanded edition of the student statutes including new chapters on professors and administrators. And by the 1570s, their responsibilities had begun to range so widely, from the selection of professors to the hearing of academic grievances against the town, to the censorship of books, that what began as a "not too . . . time consuming" office, in the words of one member of the Maggior Consiglio, became a full-time job.[24]

Once the Italian governments had reorganized the administrative structure and external aspect of the universities they then freed them once and for all from the financial preoccupations that had plagued their medieval predecessors. They supplemented student payments and duties on goods for the first time by regular government subsidies and ordered detailed records to be submitted for annual approval.[25] They made sure that increases in university outlays kept up at least in some measure with rising inflation. That of Naples, for example, raised its contribution from 900 to 4,000 ducats between 1519 and 1612. And once they were able to guarantee, as did Venice during most of the sixteenth century, that salaries would be paid on time, they could promise students that highly qualified professors would be willing to remain for more than a few years.[26]

The Italian governments then went about translating the latest intellectual developments into still more educational reforms. They made "humanities" a fully-fledged branch of higher study by creating chairs expressly for it instead of expecting Carlo Sigonio and Francesco Robortello and others to teach from the chairs of "rhetoric" and "poetry" in which it had first entered the curriculum.[27] They transformed the law faculties by hiring the best professors of the latest school of humanist jurisprudence, that of Andrea Alciato. They instituted new chairs on Justinian's *Digest* whose incumbents could make use of Lelio Torelli's recent edition of the Tuscan manuscript.[28] They transformed the medical faculties by establishing new professorships of anatomy, mathematics, chemistry, and natural history and by creating anatomical theaters and gardens of simples. And it was by just such curricular innovations in the local university, except in the faculty of law, that at least one city, Messina, managed to command an important part of the Italian intellectual landscape for the first time.[29]

Finally, the Italian governments conceived of remedies to counter what they perceived to be the most serious recent threat of all to the vitality of the universities in their states—namely, competition from other universities in Italy and in transalpine Europe. They introduced legislation excluding persons with "foreign" degrees from professional practice.[30] And when practically every government in Europe had been caught up in the race to protectionism, the Italian governments began a campaign to bring back the transalpine students. That of Tuscany, for example, sent diplomats to Spain to request recognition for Pisan degrees equal to that already granted to those of Bologna, Rome, and Naples. When their task was complicated by confessional divergences, particularly after

Pius IV, on the heels of the counter Reformation, began enforcing the requirements for attestations of orthodoxy in the granting of degrees Church-accredited institutions (i.e., most Italian universities), the Tuscan government responded by simply converting the Sapienza at Siena from a student hostel into a separate body for German students with princely recommendations. The government of Urbino sought to obtain for the local college of physicians the ability to confer degrees on all transalpine students by the ancient privilege of the Counts Palatine. And the Venetian government created a separate college, the Collegio Veneto, for awarding degrees *auctoritate Veneta*.[31]

As much to remain abreast of cultural affairs as to maintain high standards, Cosimo I himself occasionally attended the University of Pisa. "He is going around to hear these doctors," a correspondent reported on one such occasion; "and he listened to them with such attention and patience, and afterwards heard the disputations and arguments."[32] That there happened to be excellent hunting in the neighborhood, as he reported to Pedro de Toledo, not listing all "the other pleasures," was all to the good. Lelio Torelli, the celebrated jurist who advised Cosimo on all university affairs, noted that "it . . . is all for the public good and contentment of the citizens to sustain the university [of Siena]."[33] His sentiments were echoed by the representative of the viceroy of Naples, who explained that "it seemed fitting that we attend to the government and good administration of [the university], without which the divine and human sciences cannot easily be served" because "his Highness is highly interested in the increase and conservation of the public affairs of this his city and Kingdom."[34]

What is more, these governments realized that university reform might guarantee to the state a continuous supply of more and more necessary technical expertise. By the early seventeenth century, some of the problems caused by the evolution of the territorial states in their charge had grown far too complex either for the citizen-statesmen of the past or for the trained bureaucrats of the present. Challenges such as the economic depression of the 1590s or the plague of 1630, or even the Thirty Years War, made the inadequacies of previous knowledge still more evident. And few of the cultural institutions of the time were capable of collecting groups of experts who could be consulted from time to time.

These governments got little help from the Renaissance academies. Most of these academies were mainly devoted to the exercise by amateurs of their humanistic skills. If they specialized in one or another of the various branches of knowledge, such as the military academies of the Venetian Terraferma and the dictionary-making Accademia della Crusca of Florence, they still turned out to be collections of amateurs just like all the others. Even the Florentine Accademia del Disegno, with its mixed membership of artists and aristocrats, was more concentrated on general cultural conviviality than on formal consultation.[35] And while the academy sponsored a lectureship in mathematics, held in the late seventeenth century by no less an authority than Vincenzo Viviani, in part in order to sustain the academy's statutory role of overseeing civil engineering projects, little actual oversight could go on in the midst of such intense efforts devoted to "the fashioning, performance, and contestation of identities" that the latest scholarship views as a major activity. Those few academies that were specifically devoted to science, such as the Accademie dei Lincei, del Cimento, and Fisico-

Matematica (in Rome, run by Giovanni Giusto Ciampini), remained so attached to the particular policies of their founders that they disappeared as soon as the latter lost interest, died, or left town.[36]

The papal government, alone among all those in Italy, turned at times to the Jesuit colleges. For example, during the course of discussions regarding one of the major engineering problems of the day, the disastrous flooding of the Ferrarese plain, it established a special chair of mathematics at the College of Ferrara with competence "in matters concerning the surveying and regulation of water, the construction of dams, and similar things."[37] Significantly, the papal government also occasionally tapped the Jesuit network for candidates to major posts at the University of Ferrara. Governments in the other states of Italy, however, could not rely on the local Jesuits for expert assistance. Most of them found the local Jesuits to be far more interested in teaching than in providing consultation in any professional subjects except theology. After all, only the Collegio Romano enjoyed the presence, however brief, of practically all the best Jesuit philosophers, scientists, and literary figures in the order, in spite of the short-lived Jesuit university of Messina. And governments could do little to encourage the best Jesuit professors to stay in their states against Jesuit policies.

Governments found that the universities, however defective, were flexible enough to supply their needs. True, reliance upon the universities for expertise was no new thing. The Medieval universities were an endless source of legal opinions; and the Renaissance ones provided emerging political concepts with philosophical justification.[38] But during the course of the seventeenth century, governments began turning to the universities more frequently than ever before. While they turned to the top professors of medicine for assistance in times of plague, they established special chairs of hydraulic engineering for aiding in decisions affecting the safety of local agricultural land and the creation of new land out of swamps.[39] The papal government used university architects and engineers to complete two of the great engineering projects of the early modern period: the draining of the Pontine marshes and the construction of the aqueduct system that remained the city's chief water supply until 1850, bringing in some 1,700 liters of fresh water per person per day from the countryside.[40] And at least one government, that of Venice, turned the consultation of the local university into a regular policy. It called upon the professors of Greek and oriental languages whenever it had to review texts for publication that its officials could not understand. It called upon the chief theologians whenever it entered into disputes with the papal hierarchy in Rome. And it depended upon the physics professors for information on how to keep the lagoon healthy and upon the mathematics professors for advice on fortifications and ballistics.[41]

Many governments were able to hire from a pan-Italian and pan-European pool of candidates in spite of stout resistance from the universities' component colleges, and, at times, from the local aristocracies to which many professors' families belonged.[42] The Tuscan grand duke was not fooled by the courtier who assured him that "the glorious reputation of your Tuscany will be preserved" only "if [Your Highness] chooses literary men from his own state."[43] And the interstate negotiations that Cosimo I put into motion, in the case of the hiring of Andrea Alciato, including complicated maneuvers to hide his own involvement, were extended and refined by his successors and emulators.[44] The Venetian Senate competed with the grand duke for the services of Galileo; and the Bolognese senate as well as the vice royal government in Sicily competed with Pisa and Venice for the services of the rest of Galileo's disciples.

Hiring policies often seemed to suggest that research was more important than teaching. In medicine, professors interested primarily in auxiliary fields such as zoology, microbiology, embryology, and the like could aspire to the two highest-prestige posts in most universities: the first chairs of practical and theoretical medicine. The Riformatori dello Studio di Padova sometimes had to ask their representatives if the candidate under consideration—in one instance, Johann Jacob Scheuchzer, naturalist at Basel—had ever had any interest in healing at all.[45] Likewise, candidates with real interests in medical fields were often hired for chairs that were not medical, but which could be stepping-stones to the higher chairs—like Giambattista Morgagni, who was almost resigned to accepting a chair in philosophy several months before his appointment to one in anatomy.[46]

In fields such as mathematics, astronomy, botany, chemistry, and others that had worked their way into the university curriculum generally by way of medicine and that, over time, had acquired a prestige all of their own, the divergence between the recognized capacities of the professors and the interests of the students was likely to be even greater. When astronomy was finally separated from mathematics at Padua with the appointment of Geminiano Montanari in 1678, this was more in consideration of developments within these two disciplines than of the possible medical uses for one or the other.[47] Later, Giovanni Poleni was appointed to teach astronomy at Padua on the basis of his new calculating machine and a book containing his descriptions of experiments with the barometer—nothing to do with medicine.[48] The work of Borelli and Lorenzo Bellini establishing the usefulness of mathematics in the study of physiology did nothing to influence the criteria by which mathematics professors were chosen. At Padua, they continued to be hired in part because of the technical needs of the Venetian Republic. Thus, Jacob Hermann, in recommending Nicolas Bernoulli of the famous Basel mathematical family, was careful to emphasize the candidate's "ability in hydraulic architecture, which he learned in Holland, so that he will be able to serve the public usefully in this important faculty."[49] And the most important criterion for choices in all fields remained the candidate's participation in European discussions and contribution of significant works to the literature. Therefore, the Venetian Resident in London, informed about the vacant chair of botany at Padua in 1718, "spoke about the matter with Signor Cavaliere Newton."[50] Only Isaac Newton's insistence, based on his conversations with Hans Sloane, that "this subject is particularly studied in Italy," ensured that the choice would eventually fall on an Italian.

This is not to say the best interests were always served, especially when promotions were awarded exclusively on the basis of favoritism, as in the case of Nicolo Calliachi, the well-paid professor of belle lettere at Padua who had privately tutored a senator, or else on the basis of seniority, as in the case of Michelangelo Molinetto, professor of anatomy at Padua, who wrote practically nothing and was so unpopular that he sometimes had only one student a year.[51]

Frequently enough, the same criteria were used in promotions as in the hiring process. Morgagni finished his career with a stipend of over 2000 ducats a year because of the productivity of his collaboration with Albrecht von Haller and other important transalpine scholars, because of the fame of his *Adversaria anatomica* (Bologna, 1706 - Padua, 1719), a treatise revising Vesalius and Fabricius ab Aquapendente, and because of the usefulness of his *De sedibus* (Venice, 1761), the first systematic work on pathology written in the eighteenth century.

Making sure the best and most innovative hires stayed put called for the granting of some privileges in regard to teaching. At Bologna, a decree was issued encouraging private lessons so the best professors could prepare their tiny groups of disciples without interference from exam-conscious students. And the professors openly applauded the officials' actions. "Bologna [has always had its] good, mediocre, and weak [professors]" wrote Marcello Malpighi, "But experience has always shown that among this group there is always a few who support Bologna and Italy."[52] While universities became the chief centers for scientific and legal speculation, administrators became some of the best apologists for exactly the same version of the new science that the most innovative Italian scientists considered to be consonant with Galileian traditions. "It might be a good idea just to say Chemistry," noted an official at Pisa concerning the prospect of introducing the new discipline into the medical program. "To call it chemical medicine," however, would be to suggest the animist and vitalist ideas of the transalpine Paracelsians, which were "damned in all the universities" of Italy in favor of a purely mechanical approach.[53] Nevertheless, noted yet another official, "the most illustrious discoveries of other countries do not take long to come to the notice of the university professors," and this, he said, "has had a considerable influence on perfecting the method used . . . in teaching."[54]

The governments of the Italian states did not force colleges to update their exams to conform to the teaching of the innovative professors; nor did they force the innovative professors to obey the decrees that they repeated, with increasing frequency, requiring them to teach from the standard texts. Instead, they devised various stratagems to ensure that students could always obtain good, approved, informal instruction in the proper texts somewhere in the university. "The public lessons in the Bò are no longer sufficient for preparing students for the doctorate," proclaimed the Riformatori dello Studio di Padova. One strategy was suggested by Lorenzo Magalotti in Florence: let the government choose a candidate who could "read well the Peripatetic school" and at the same time "be otherwise a free [i.e., anti-Aristotelian] philosopher"—in his spare time.[55] Needless to say, the search for such a candidate was a long one.

The Riformatori dello Studio di Padova suggested a simpler alternative: they ordered the yearly selection of four "puntisti" or coaches to help students with exams.[56] The senate of Bologna, in the same spirit, inflated the number of lower-rank positions available for instructors with few ambitions and a knowledge of the standard texts. These new appointments were the chief causes for the growth of the faculties of law and medicine during the course of the seventeenth century, shown in Table 4.1. Thus, at least part of the university faculty was always de-

Table 4.1: Faculties of Law and Medicine at Bologna, 1585-1677

	1585	1600	1630	1633	1656	1677
Law	30	39	38	43	63	79
Medicine	71	59	54	52	65	75

Source: Umberto Dallari, *I rotuli dei lettori legisti ed artisti dello studio bolognese* vol. 2 (Bologna: 1889).

voted to the texts that the colleges of medicine and law continued to require in their examinations. These and other universities also began to enforce attendance at lessons by requiring professors to keep records.

Exactly what effect government and college policies may have had on the students themselves is difficult to say. Those students who were more interested in passing exams than in understanding the world around them may have considered attendance at lessons, especially by the innovators, to be an expensive and time-consuming luxury. Who could blame them for experiencing difficulty taking notes on what they heard, even if they were impressed by brilliant displays of scholarly and scientific erudition?[57] When they realized that many professors did not provide the familiarity they needed with the ancient texts, some of them did the most obvious thing—they abandoned the university, sought tutors from among the many university graduates that university expansion had placed in all the small towns of Italy and returned to the university only to take exams. Others never returned. In the late sixteenth century these students were usually replaced by new matriculants, so totals were not affected. After the depression and plague

Table 4.2: Degrees Awarded at the University of Pisa, 1610-1699

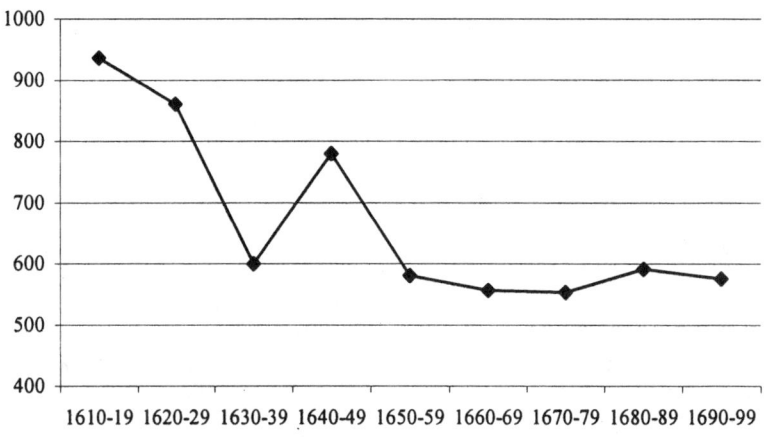

Source: Giuliana Volpi, "Lineamenti per uno studio sull'Università di Pisa nel XVII secolo," *Scritti in onore di Dante Gaeta* (Milan: Giuffré, 1984).

Table 4.3: Matriculants at the University of Bologna, 1550-1699

1550-99	2372
1600-49	2492
1650-99	2935

Source: Gian Paolo Brizzi, "Matricole ed effettivi," in Brizzi and Antonio Ivan Pini, *Studenti e università degli studenti dal XII al XIX secolo* (Bologna: Ist. per la storia dell'università, 1988).

of 1620-30 the decline began; and by the 1640s, the number of degrees awarded was at an all-time low.[58] Consider the figures in Table 4.2, concerning the University of Pisa. However, figures on the numbers of degrees awarded are no more conclusive, in constructing a realistic picture of student demand, than are figures on enrollment. Those who came only in order to benefit from student privileges and exemptions were just as unlikely to appreciate the improvements in university culture that were being made at this time as were graduates who never attended classes. Nor were students the only ones who might benefit from university teaching, if the pattern already revealed in a famous late sixteenth-century case is any indication. It is said that even a semi-ignorant miller in Friuli was capable of picking up some of the more unorthodox ideas then being distributed from the podium at the University of Padua.[59] Why not also in the seventeenth century?

The number of students who found in the universities a stimulating environment for study and contemplation may well have grown during the course of the century, helping to constitute the growth spurt that would set the tone for the following century. Consider the figures for the University of Bologna in Table 4.3. At Padua, as Table 4.4 shows, enrollments oscillated extraordinarily around the 3,000 and 3,500 student mark for three-year periods. Even at Pisa, as Table 4.2 showed, the number of degrees awarded eventually leveled off after the decline in the 20s and 30s, although it never really recovered.

The most important consequence of the governments' tolerance and the colleges' rigidity was the unexpected emergence of a new kind of university. Neither the governments nor the colleges succeeded in imposing an intellectual hegemony through their policies toward teaching. University officials did not even fully succeed in their projects for using the universities as centers for humanist patronage. The implementation of such projects was always conditioned by compromises and arrangements made in response to the conditions that from time to time arose. Yet the universities acquired a definite physiognomy during the early modern period, one that was entirely independent of the ideas of any one group. This physiognomy, complex and little understood, might be called a dual university system, since it ensured that one part of the university was always devoted to scientific innovation and the other to avoiding it.

Yet another consequence of the attitudes of government and colleges was a certain confusion in the emergence of a distinct profession of university instructor. At first, attempts by the Galileians to install their ideas permanently in the

Table 4.4: Matriculants at the University of Padua, 1633-97

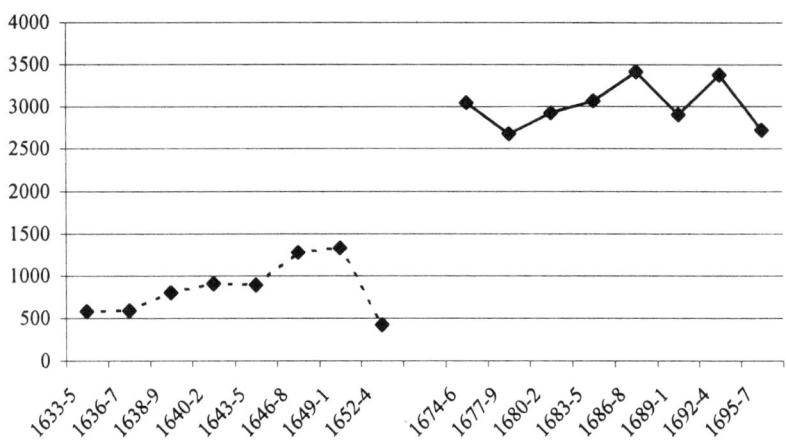

Source: Mario Saibante, *Studi di economia, statistica e sociologia, 1924-55* (Rome, 1959). Values for 1633-54 are hypothetical.

universities seemed likely to destroy forever the prestige of their non-innovative colleagues. The production of scientific discoveries seemed likely to gain a permanent place in all routine decisions about hiring and promotion. But dissent among the innovators themselves gave governments interested in evaluating contributions a difficult job.[60] What to do, for instance, when mechanicists adamantly refused to endorse animists, although both agreed on the fundamental purposes of research? Moreover, some members undermined their own position by insisting that so-called scientific discoveries constituted nothing more than the rediscovery of knowledge that had existed at some time in the past. Thus, what had been a defense tactic of the atomists, in asserting a Democritean or Epicurean heritage, over the long term could work to their disadvantage.

Meanwhile, members of the non-innovative part of the university proved to be far more tenacious than their opponents supposed. They attempted to beat their innovative colleagues at their own game—scientific publication—by writing elaborate treatises in defense of the authors upon which the traditional curriculum was based and thereby earning the names "Aristotelian," "Galenist," or the like. They attempted to win away from their colleagues credit for a new method of proving scientific conclusions by using visual tests similar to experiments yet aimed at a different purpose—that of demonstrating the veracity of the ancient authors' precepts.[61] And to ensure that these tactics should not wear down the distinctions too much between themselves and others, they enlisted the support of ecclesiastical authorities who did not have time to revise outdated views about the necessity for philosophical homogeneity, as in the atheist trials in late seventeenth-century Naples. They enlisted the support of heads of government, too, who believed the appeasement of ecclesiastical authorities could serve a political point—as did Cosimo III of Tuscany when he prohibited Cartesianism at Pisa in the same period.[62] So successful, indeed, were their attacks that the innovators were sometimes driven to defend themselves by asserting that they were simply abiding, in a novel way, by the very statutes that they had been ignoring for years.[63] Thus, as late as the 1690s at least some instructors could claim that the definition of their profession should take into account the desires of students to obtain knowledge of the ancient texts. At times, they even elevated this idea to a principle of international cultural politics: "We must . . . sustain our ancient medicine [and] our ancient philosophy," wrote one professor to the Venetian government, "in order to keep our schools different from [those of the transalpine states], so that when the transalpine students come they will . . . see the differences between their schools and our own."[64] As late as 1731, the Newtonian physics professor Giuseppe Roma at the University of Turin was accused of "impressing dangerous things . . . on the minds of the young."[65]

Professors often found that their new ideas might be better accepted outside than inside the universities. Their efforts to formulate a professional identity within autonomous institutions contributed to a new academic movement between the seventeenth and the eighteenth centuries, distinct from the pattern set in the Renaissance. While the Accademia del Cimento in Florence reflected the patronage strategies of the Medici dynasty, the Accademia degli Investiganti in Naples reflected opposition by professors Tommaso Cornelio and Francesco d'Andrea against the medical establishment in the University of Naples. In Siena, the Accademia dei Fisiocritici was founded by Pirro Maria Gabrielli, lecturer in medicine and botany. In Bologna as in other great university centers, private informal academies, at the homes of Geminiano Montanari, Sebastiano Melli, and Giacomo Sandri, were often the direct extension of university teaching. Meanwhile, the Accademie della Traccia and degli Inquieti invited these and other university professors to discuss research interests in an appropriate setting. On these foundations, Luigi Ferdinando Marsili would build the Bolognese Institute in 1713, with the explicit purpose of complementing university education by practical instruction in the sciences.[66]

Ideas for university reform between the seventeenth and eighteenth centuries were generally well in advance of what could actually be accomplished. The new chancellor of the University of Bologna in 1687, Anton Felice Marsili, was neither the first nor the last to declaim against "the dreadful catastrophe" he saw around him. "Disciples are disappearing, podiums are empty, the masters are mute." To put an end to the same features Federico Cesi had noticed in the beginning of the century, and a few more that he believed had ruined the university's reputation among the foreigners who stayed away to attend universities in their own states, he suggested removing the root cause of all the rest. By canceling the privileges of the professional colleges, especially in the hiring process, he hoped to put the task of updating the faculties on a modern footing once and for all. His programs were no more successful than were those of his brother, Luigi Ferdinando, who proposed a long list of new appointments in modern subjects before turning his attention instead to founding the Institute.[67]

Scipione Maffei, presenting reform proposals first to the Riformatori dello Studio di Padova and then to the Duke of Savoy, was even more ambitious.[68] And with good reason, he maintained. How could anyone expect a university to make a favorable impression on prospective students where there were no less than five full professors of the books of Avicenna? Who could imagine significant progress being made where ten or twelve professors were supposed to be explicating the books of Aristotle? Let all such chairs be replaced by chairs in the modern scientific disciplines. That the professors who held the chairs in question often did not teach the materials announced by such quaint titles, meant little to Maffei, who had never attended a university. Like the brothers Marsili, he hoped for nothing less than the introduction of professional scholarly standards into all college education and the replacement of benign tolerance with vigilant surveillance by the tribunal of scientists and literary figures, in association with government bureaucrats.

In subsequent decades, those responsible for putting such projects into action responded to the new challenge by putting the finishing touches on the institutions that their predecessors had created in the last two centuries. The Duke of Savoy created a governing board for the University of Turin along the lines of the Riformatori dello Studio di Padova. The Riformatori and and their counterparts elsewhere appointed new professors with new interests. Some of them even created new structures for experimental physics, chemistry, natural history and astronomy, including new professorships and staff to run them.[69] They set up chairs of surgery to provide medical schools with expertise in what was fast becoming a legitimate scientific specialty.[70] They never dared to remove the benefits their states received from the dual university system.

Even after mid-century, a system divided between tradition-bound teaching and advanced research continued to haunt the minds of university officials. "Public lessons are mainly for mediocre students who want to enter the professions," noted one. "The private ones are for sublime geniuses."[71] All agreed that the existing arrangement served the best purposes of government. "The aim of the university," noted another official, "[is] . . . first of all to distribute the sciences," which he evidently considered to be static or slowly changing sets of principles, "throughout [the state], especially those that will permit . . . subjects to exercise ably all types of professions. . . ; [and] second[ly] to keep and preserve in this [state] the most noble disciplines"—in other words, the most advanced investigations of the specialists—raising those disciplines "to ever greater levels of excellence and perfection and, if possible, to widen their confines." And the way to do this, he remarked, was to "give fitting emoluments to exalted minds both to reward them for their illustrious accomplishments and to encourage them to make ever-greater progress and discoveries for the benefit and ornament of humanity." By appointments, in other words, and not by legislation. Exactly the same practice as in previous centuries.[72]

The dual university system of the early modern period did not survive the second major challenge. Political crises in many of the Italian states at mid-century turned the attention of government officials toward the more radical ideas of reformers such as Bernardo Tanucci and Bartolomeo Intieri in Naples, and Pietro and Alessandro Verri in Milan. These officials were less concerned than their predecessors had been to tread softly around the professional corporations—whose power over exams they severely circumscribed in Naples and entirely abolished in Milan.[73] Moreover, it was no longer enough to give the appearance of progress by intervening so cautiously in the university structure, adding new professorships and institutes. With benign tolerance giving way to enlightened action, they set about to reform the whole curriculum according to new standards. For the first time, the actual matters being taught in the classroom were placed under careful scrutiny. "Bacon, Locke, Condillac, and Charles Bonnet, and not [the Italian theologian Sigismondo] Gerdil," insisted the Austrian official in Milan, Wenzel Anton Kaunitz, "must be the masters," calling for the fusion of Illuminismo with the European Enlightenment.[74]

In the course of comprehensive plans for bureaucratic reorganization encouraged by these reformers, the Italian governments adopted a new approach to university reform. For the first time, they enacted legislation to abolish the distinctions between the scientific specializations in which they desired the aid of the professors' expertise and the disciplines taught in the university. That of Venice installed a new professorship of agriculture at Padua just when grain prices were caught in spiraling inflation, and one in veterinary medicine as more theorists began to insist on the benefits of agricultural diversification. It installed one in Venetian law just when it planned to correct its relationship to the provinces and one in public ecclesiastical law just when it was about to defend massive ecclesiastical expropriations.[75] That of Milan separated the faculties at Pavia to make them more representative of the disciplines taught—theology from philosophy, philosophy from medicine—and reorganized them according to the Austrian four-faculty model. It then overhauled the medical school to make it supply everything necessary for public health including obstetrics.[76] And it overhauled the theological school to make it supply everything necessary for the tutelage of religion that it had recently taken over from Rome, including the unorthodox ideas of Cornelius Jansen, which had hitherto been, at least formally, avoided. The government of Naples introduced the first chair of political economy to help itself out of a devastating trade imbalance with surrounding states; and those of Modena and Bologna soon followed. More radical yet, it broke a 500-year-old tradition by commanding that lessons be given in the vernacular instead of in Latin.[77] Finally, at least some governments, such as those of Tuscany and Venice, considered the possibility of further unifying curriculum and research by promoting a new literary genre that had recently evolved in transalpine Europe—the university journal. And in doing so they implicitly agreed with the university reformer who noted that "making one's knowledge public by writings and printed works" was "far more noble" than "diffusing it . . . to student listeners," advancing the profession of university instructor one more evolutionary step.[78]

While something was indeed lost in the new cultural policies, much was also gained. The Austrian government's sponsorship of well-equipped laboratories and copiously furnished museums provided the context for the activity of Alessandro Volta, Spallanzani and Antonio Scarpa at Pavia. The same went for the Habsburg-Lorraine government, which sponsored Felice Fontana and Giuseppe Fabbroni at Florence, as well as the Papal government, which spared no pains to ensure that Luigi Galvani and Leopoldo Marcantonio Caldani at the University of Bologna had everything they needed to pursue their work. Venice was not far behind in maintaining Marco Carburi and Giuseppe Toaldo at the University of Padua. And in the Savoy state, military necessities added impetus to government sponsorship of culture so that Louis Lagrange, Francesco Domenico Michelotti, and eventually Amedeo Avogadro could keep the scientific and technological faculty at the university at least on par with the rest of Europe. An "Italian revolution," Spallanzani's publisher called it, both in the sciences and in politics, was under way.[79]

However much the eighteenth-century governments may have disrupted university life by their authoritarian intervention, they helped defeat negative market tendencies that discouraged innovation and research. Instead, by making universities important parts of an overall educational program, they succeeded in bringing back the students. At Padua alone enrollment doubled between 1760 and 1782.[80] And they finished the task of introducing usefulness as a criterion in the evaluation of academic knowledge. At the same time, they did their best to provide an atmosphere favorable to cultural exploration—scientific and literary—in the midst of social conflict and economic turmoil. In part due to their efforts, Italy took its place once again at the center of cultural debates. By the late eighteenth century, public perceptions began to turn against the enemies of innovation; and erstwhile enemies of innovation adopted once again the tactics of the innovators themselves, as we shall see.

Notes

1. Federico Cesi, "Del naturale desiderio di sapere" in Ezio Raimondi, ed., *Narratori e trattatisti del Seicento* (Milan-Naples: Ricciardi, 1960).
2. Marina Roggero, *"Insegnar lettere": ricerche di storia dell'istruzione in età moderna* (Alessandria: Edizioni dell'Orso, 1992), 49. See, however, Paolo Prodi, "L'università nell'età confessionale tra Chiesa e stati: secoli XV-XVII," *Annali dell'Istituto storico italo-germanico in Trento* 17 (1991): 11-23.
3. Steven Shapin, *The Scientific Revolution* (Chicago: University of Chicago Press, 1996).
4. Some of the more recent work on university faculties includes Emanuele Conte, *I maestri della Sapienza dal 1514 al 1787,* 2 vols. (Rome: Edizioni dell'Ateneo, 1991); Francesco Raspadori, ed., *Maestri di medicina ed arti dell'Università di Ferrara, 1391-1950* (Florence: Olschki, 1991); Lucia Rossetti, ed., *Rapporti tra le università di Padova e Bologna* (Trieste: Edizioni Lint, 1987); Gian Paolo Brizzi and Angelo Varni, eds., *L'università in Italia fra età moderna e contempoanea. Aspetti e momenti* (Bologna: Clueb, 1991).
5. BNM, mss. lat. cl. 8:157 (=2734), cc. nn.
6. Compare the fairly balanced resumé of recent work given by Roy Porter, in Hilde de Ridder-Symoens, ed., *A History of the University in Europe, vol. 2: Universities in Early Modern Europe* (Cambridge: Cambridge University Press, 1996), chap. 13, to the more traditional account of Olaf Pedersen in the same volume, chap. 11. A good recent resumé of recent work on Italian universities is Marina Roggero, "Le università in epoca moderna. Ricerche e prospettive," Luciana Sitran Rea, ed., *La storia delle università italiane. Archivi, fonti, indirizzi di ricerca. Atti del convegno, Padova, 27-29 ottobre, 1994* (Trieste: Lint, 1996), 311-34. In addition, Boguslaw Lesnodorski, "Les universités au Siècle des Lumières," in *Les universités europèens du XVIème au XVIIIème siècle* (Geneva: Droz, 1967), 143-60; Helmut Coing, "Des juristische Vorlesungsprogramm der

Universität Padua im 17. und 18. Jahrh.," in *Studi in onore di Edoardo Volterra*, 6 vols (Milan: A. Giuffré, 1971), 4: 179-95; Sandro De Bernardin, "La politica culturale della Repubblica di Venezia e l'Università di Padova nel secolo diciassettesimo," *Studi veneziani* 16 (1974): 443-502; Elsa Mango Tomei, *Gli studenti dell'Università di Pisa sotto il regime Granducale* (Pisa: Pacini, 1976), 61-64; Maria Rosa Di Simone, *La "Sapienza" romana nel Settecento* (Rome: Edizioni dell'Ateneo, 1980), 19-28; Giuseppe Ricuperati, "Università e scuola in Italia," in *La letteratura italiana, I: Il letterato e le istituzioni* (Turin: Einaudi, 1982), 983-1007.

7. The practice was well entrenched by the early seventeenth century, notes Federico Cesi, "Del naturale desiderio di sapere," 50. Anthony Grafton and Lisa Jardine, in *From Humanism to the Humanities* (Cambridge, Mass.: Harvard University Press, 1986), 61-3, trace the change among a few professors in the humanities already to the late Quattrocento.

8. Nancy G. Siraisi, "Renaissance Commentaries on Avicenna's Canon," *History of Universities* 4 (1984): 63, her translation. The other references are to Antonio Poppi, "Introduzione," in Pietro Pomponazzi, *Corsi inediti dell'insegnamento padovano, I: Super libello de substantia orbis expositio et quaestiones quattuor (1507)*, ed. A Poppi (Padua: Antenore, 1966), xv; Giustina Simionato, "Significato e contenuto delle "lectiones' inedite di logica di Bernardino Tomitano," *Quaderni per la Storia dell'Università di Padova*, 6 (1973): 119, as well as Maria Rosa Davi, *Bernardino Tomitano, filosofo, medico e letterato (1517-76). Profilo biografico e critico* (Trieste: Lint, 1995), chap. 2.

9. Christopher Lewis, *The Merton Tradition in Kinematics in the Late Sixteenth and Early Seventeenth Centuries* (Padua: Antenore, 1980), 7; William Wallace, *Galileo and his Sources* (Princeton: Princeton University Press, 1984), 93, 228.

10. For example, Ugo Baldini, "La scuola galileiana" and "L'attività scientifica del primo Settecento," in *Storia d'Italia*, Annali 3: *Scienza e tecnica* (Turin: Einaudi, 1980), 383-551. In addition, Corrado Dollo, "Fra tradizione e innovazione. L'insegnemtneo messinese della medicina e delle scienze nei secoli XVI e XVII," *Annali di storia delle università italiane* 2 (1998): 107-22.

11. This and the following are from Bellini's letter, 17 November 1689, in Howard B. Adelmann (ed.), *The Correspondence of Marcello Malpighi*, 5 vols (Ithaca, N.Y.: Cornell University Press, 1975), 5: 1553.

12. I discuss these issues more fully in the first chapter of my *Science, Politics and Society in Eighteenth-Century Italy: The "Giornale de' Letterati" and its World* (New York: Garland, 1991).

13. Howard B. Adelmann, *Marcello Malpighi and the Evolution of Embryology*, 5 vols. (Ithaca: Cornell University Press, 1966), 1: 542-43; Giambattista Morgagni, *Opera postuma*, 1: *Le autobiografie*, ed. Adalberto Pazzini (Rome: Istituto di storia della medicina dell'Università, 1964), 18-19.

14. Davide Giordano, "Alessandro Knips Maccope," *Rivista di storia delle scienze mediche e naturali* 27 (1936): 127.

15. Bianca Betto, *I Collegi dei notai, dei giudici dei medici e dei nobili in Treviso, sec. 13-16*, Deputazione di Storia Patria per le Venezie, Miscellanea di Studi e Memorie, vol. 19 (Venice: Deputazione Editrice, 1981), 239; Charles B. Schmitt, "Science in the Italian Universities in the Sixteenth and Early Seventeenth Centuries," in *The Emergence of Science in Western Europe*, ed. Maurice Crosland (London: Macmillan, 1976), 35-56. Consider also Alessandro Gibba, "Francesco de' Vieri (1534-91) and his teaching at the University of Pisa," *History of Universities* 14 (1995-6): 143-55.

16. Anthony Grafton, "Teacher, Text, and Pupil in the Renaissance Classroom: A Case Study from a Parisian C ollege," *History of Universities* 1 (1981): 37-40.

17. Gabriele Baroncini, "La filosofia naturale nello Studio bolognese (1650-1750)," in *Scienza e letteratura nella cultura italiana del Settecento*, ed. Renzo Cremante and Walter Tega (Bologna: Il Mulino, 1984), 289-90.

18. R. Burr Litchfield, "Ufficiali ed uffici a Firenze sotto il granducato mediceo," Elena Fasano Guarini, ed., *Potere e società negli stati regionali italiani del "500 e "600* (Bologna, Il Mulino, 1970), 133-151; Danilo Marrara, *Riseduti e nobiltà. Profilo storico-istituzionale di un oligarchia toscano nei secoli 16-18* (Pisa: Pacini, 1976), 135; Betto, *Il collegio dei notai, dei giudici, dei medici e dei nobili in Treviso*, 133-50; 223-44.

19. Two easily consultable examples: Virginia Cordero di Montezemolo and Ugo Gualazzini, eds., *Corpus Statutorum Almi Studij Parmensis (saec. XV)* (Milan: Giuffrè, 1946), 52; Carlo Malagola, ed., *Statuti delle università e dei collegi dello studio bolognese* (Bologna: Il Mediterraneo, 1888), 385.

20. Robert Palmer's statement that "the colleges of arts and medicine were in fact the civic colleges of physicians" seems inexact, in "Physicians and the State in Post-Medieval Italy," Andrew W. Russell, ed., *The Town and State Physician in Europe from the Middle Ages to the Enlightenment* (Wolfenbüttel: Herzog August Bibliothek, 1981), 50. The best discussion of the documents, which contain considerable lacunae for the important cases of Padua and Bologna, is in Bianca Betto, *Il collegio dei notai, dei giudici, dei medici e dei nobili in Treviso*, 234-37.

21. Stefano De Rosa, "Studi sull'Università di Pisa: II: La riforma e il paradosso," *History of Universities* 3 (1983): 107; the statute of 1544, chap. 53, is quoted by Giovanni Cascio Pratilli, *L'Università e il principe: Gli studi di Siena e Pisa tra rinascimento e riforma* (Florence: Olschki, 1975), 132; and the statement by college of law in 1573 is quoted in Rodolfo Del Gratta, "Spigolature storiche sull'Università di Pisa," *Università e società nei secoli XII-XVI* (Pistoia: Centro Italiano di Studi di Storia e d'Arte, 1982), 297.

22. Decree of the Tuscan government, quoted in Danilo Marrara *Lo Studio di Siena nelle riforme del Granduca Ferdinando I* (Milan: Giuffré, 1970), 163. Similarly, the Bolognese Senate, in 1591, *Memorie e documenti per la storia dell'Università di Pavia*, 3 vols. (Pavia: 1877-78, repr. Bologna: Forni, 1970), 2: 20.

23. ASF, *Archivio Mediceo del Principato*, filza 2, fol. 492r, 21 November 1542, Cosimo I to Alfonso II d' Ávalos de Aquino, (Marqués de Vasto) [Doc. 7432 in MAP]: "Anchor che la E. V. sia occupata totalmente nelle cose della guerra, non intendo di [cancelled; gravarla] ricercarla de una cosa che riguarda piu el tempo della pace . . . tal mio desiderio in questo et ^che^ [cancelled: come] havendo io ^qualche anno^ [cancelled: più anni] fa desiderato [illegible through ink loss] per molte cause di indirizzarsi a rinnovar lo studio in questa città di Pisa. . . . Però havendo fra gli altri doctori che sono necessarii a tale impresa mancamento di un medico famoso et excellente, et sapendo che Maestro Matteo da Corte esser tale, et che verbi gratia per essere vassalo di S. M.za gli puo comandare che lui per non esser obligato ad altri puo facilmente accetare il carico di venire a leggere in tale studio. Nel quale non solo non hanno mai ricusato, anzi desiderato molto leggere tutti li primi homini di antiquità et di fama non punto inferiore ad alchuno altro di Italia, con quelle conditioni che seranno honorevoli et ragionevoli et che la S. V. medesima giudicherà esser convenienti . . . certificandola che fra molti et molti ^comodi^[cancelled: servitii] et piaceri che la si può fare, questo sara reputato da me grandissimo [cancelled; non lo reputarò de è uno delli maggiori]. Et oltre che à M.ro Matteo sara offerto quanto sara permesso secondo el consueto delli pagamenti [cancelled: salarii] di Firenze et non di quelli di Padova." Concerning the aftermath of the Italian Wars, Eric Cochrane, *Italy, 1530-1630*, ed. Julius Kirshner (London: Longmans, 1988), chap. 2: "The New Political Order."

24. The quote is from François Dupuigrenet Desroussilles, "L'Università di Padova dal 1405," 642. So well regarded was the office of Riformatori that candidates were often

catapulted from it into the dogeship, notes Sandro De Bernardin, "I Riformatori dello Studio di Padova: Indirizzi di politica culturale nell'università di Padova," *Storia della cultura veneta*, vol. 4: *Il Seicento*, part 2 (Vicenza: Neri Pozza, 1984), 63-64. Similar policies in other universities are analyzed in: Maria Claudia Toniolo Fascione, "Aspetti di politica culturale e scolastica nell'età di Cosimo I: l'istituzione del collegio della Sapienza di Pisa," *Bollettino storico pisano* 49 (1980): 61-86; Stefano De Rosa, *Una biblioteca universitaria del secondo Seicento: la libreria di Sapienza dello studio Pisano, 1666-1700* (Florence: Olschki, 1983); Alessandro D'Alessandro, "Materiali per la storia dello Studium di Parma, 1545-1622," *Università, principe, gesuiti. La politica farnesiana dell'istruzione a Parma e Piacenza, 1545-1622* (Rome: Bulzoni, 1980), 21; Giuseppe Ermini, *Storia dell'Università di Perugia*, 2nd ed., 2 vols. (Florence: Olschki, 1971), 1: 210; Nicola Spano, *Storia dell'Università di Roma* (Rome: Casa Editrice "Mediterraneo," 1935), 15-20; Mario Caravale and Alberto Caracciolo, *Lo stato pontificio da Martino V a Pio IX* (Turin: U.T.E.T., 1978), 310-35.

25. A late example concerning the university of Mantua is in ASF, *Principato*, filza 6109, unnumbered folios, dated 18 October 1625, Ferdinando I Gonzaga to Caterina de' Medici [MAP document no. 7129]: "Ancorchè io non habbia per anco potuto ben vedere le consideratione che vengono addotte dai mercanti della lana circa la nuova impositione per lo Studio, onde non posso liberamente dichiarare la mia volontà; nondimeno, dovendosi fare qualche provisione per l'iminente bisogno di esso Studio, che ricerca l'appigliarsi a qualche partito, parmi che per hora sia il più spedito e men dannoso e sensibile l'aggiungere il sesino per libra alla carne per tutto lo stato."

26. The financial records pertaining to Padua are contained in *Documenti finanziari della Repubblica di Venezia, ser. 2: Bilanci generali*, 3 vols. (Venice: R. Commissione per la Pubblicazione dei Documenti Finanziari della Repubblica di Venezia, 1903-12). In addition, F. Dupuigrenet Desroussilles, "L'Università di Padova dal 1405," 639; N. Spano, *Storia dell'Università di Roma*, 20; Nino Cortese, "L'Età Spagnola," *Storia dell'Università di Napoli*, eds. N. Cortese and Michelangelo Schipa (Naples: Ricciardi, 1924), 224.

27. Umberto Dallari, *Rotuli dei lettori dello Studio di Bologna*, 5 vols. (Bologna: Merlani; later, Deputazione di Storia Patria, 1888-1929), 1: 107; 2 (1889): 190; Angelo Fabroni, *Historiae Academiae Pisanae*, 3 vols. (Pisa: Mugnai, 1792), 2: 471; Carlo De Frede, *I lettori di umanità nello studio di Napoli durante il Rinascimento* (Naples: L'Arte Tipografica, 1960), 193; Luigi Tondo, "Celio Calcagnini: l'uomo e l'umanista," Patrizia Castelli, ed., *'In supreme dignitatis.' Per la storia dell'Università di Ferrara, 1391-1991* (Florence: Olschki, 1995), 173-84. Note also Paul Grendler, "Five Italian Occurrences of 'Humanist,'" *Renaissance Quarterly* 20 (1967): 317-25.

28. G. Cascio Pratilli, *L'Università e il Principe*, 93, 147, 153; Paul Oskar Kristeller, "The University of Bologna and the Renaissance," *Studi e memorie per la storia dell'Università di Bologna* 1 (1956): 313-23; Maria Carla Zorzoli, "Interventi dei duchi e del senato di Milano per l'Università di Pavia," *Studi senesi* 102 (1980): 128-49; Alberto Lupano, "L'insegnamento ed il soggiorno ferrarese del giurista Aimone Cravetta," '*In supreme dignitatis,*' 505-24. Concerning Messina, Daniela Novarese, "'Che i legisti debbano fondare le lectioni loro sopra Bartolo,'" *Annali di storia delle università italiane* 2 (1998): 73-83.

29. Rosario Moscheo, "Scienza e cultura a Messina fra Cinque e Seicento. Vicende e dispersione dei manoscritti autografi di Francesco Maurolico," *La rivolta di Messina, 1674-8 e il mondo mediterraneo nella seconda metà del Seicento*, Convegno Storico Internazionale, Messina 10-12 ottobre, 1975, ed. Saverio di Bella (Naples: L.P.E., 1978), 435-74. The chief authority on modifications in science curricula is still Charles B. Schmitt, whose various papers on the subject are collected in *The Aristotelian Tradition and Renaissance Universities* (London: Variorum Reprints, 1984).

30. Exemplary is the decree of Francesco Sforza in Milan, 7 October 1522, *Memorie e documenti per la storia dell'Università di Pavia*, 2: 17.

31. S. De Bernardin, "La politica culturale del governo veneziano," 453.

32. ASF, *Archivio Mediceo del Principato*, filza 6, fol. 411r, 11 December 1545, Cosimo I to Pedro de Toledo: "La duchessa [Eleonora di Toledo] et io con li figlioli [Maria, Francesco, Isabella, Giovanni, Lucrezia] stiamo bene et andian [andiamo] godendo insieme col Sig.r Don Luigi [Luis de Toledo] questa bella stantia di Pisa, che è un piacer a starci, sì per lo Studio, il quale è floridissimo, come per le caccie commode et di piacere che ci si trovano, che non ci mancha altro che la presentia della Ecc.tia V." Filza 1171, fol. 283r, 30 January 1544, Vincenzo Ridolfi to Pier Francesco Riccio [Doc. 7059 in MAP]: "Seguendo S. Ex. [Cosimo I] d'andare a udire questi dottori, hier l'altro fu alla lettione del Remigio [Migliorati dal Borgo], et del [cancelled: Decano] Brando [Branda Porro], et hierj a quelle del Decano, et del Lapino [Antonio Lapini], et con tale attentione et patientia gli ha uditi, et poi sentiti disputare, et arguire, che più la S. V. non si potrebbe immaginare. Hoggi par che si sia fatta la vacatione, respetto alla notomia, havendo jl Vesalio [Andreas Vesalius] cominciato a vedere et leggere quelle cose delli ossi, de' quali non si è potuto fare lo schieleto intero, perchè il cadavero che venne di costà [Firenze] havea rotto non so che costole, sicchè el mal suo non fu pleura come qua fu avisato. Et doppo questo si anderà più innanzj et preparasi perchè questa vuol che sia l'ultima di fare cose grande et farassi di più corpi, poichè de' subretti non manca, così di ^più^ huomini, come d'altri animali."

33. Letter to Cosimo I, 1574, quoted in G. Cascio Pratilli, *L'Università e il principe*, 181.

34. Nino Cortese, "L'Età Spagnola," 272, decree of 1507 by Giovanni d'Aragona, luogotenente.

35. Giuseppe Olmi, "'In essercitio universale di contemplatione, e prattica': Federico Cesi e i Lincei," Laetitia Boehm, ed., *Università, accademie e società scientifiche in Italia e in Germania dal Cinquecento al Settecento*, Annali dell'Istituto Storico Italo-Germanico in Trento, Quaderno 9 (Bologna: Il Mulino, 1981), 173, shows that members gave "lessons" in Platonic philosophy, metaphysics, botany, mathematics, astronomy and history; J. R. Hale, "Military Academies on the Venetian Terraferma in the Early Seventeenth Century," *Studi veneziani* 15 (1973): 273-95. In general, Eric Cochrane, "The Renaissance Academies in their Italian and European Setting," *The Fairest Flower: The Emergence of Linguistic National Consciousness in Renaissance Europe* (Florence: Accademia della Crusca, 1985), 21-39.

36. William E. Knowles Middleton, "Science in Rome, 1675-1700, and the Accademia Fisico-Matematica of Giovanni Giusto Ciampini," *British Journal for the History of Science* 8 (1975): 138-54. Concerning the Accademia del Disegno, Karen-Edis Barzman, *The Florentine Academy and the Early Modern State. The Discipline of Disegno* (Cambridge: Cambridge University Press, 2000), 19.

37. Ferrante Borsetti, *Historia almi Ferrariae gymnasii*, 2 vols. (Ferrara: Pomatelli, 1737), 1: 310, from the papal decree: "Il bisogno che ha questo Pubblico di persone, che ne siano sufficentemente erudite affinchè in congiuntura di Livellazioni, o regolamenti di acque, costruzioni di arginature, e simili altre occorrenze, possano servire con profitto alla patria [è purtroppo frequente]." Consider also Luigi Pepe, "La crisi dell'insegnamento scientifico dei Gesuiti a Ferrara e l'inizio dell'attività didattica di Teodoro Bonati," *'In supreme dignitatis,'* 61-74. In addition, for what follows, Daniela Novarese, *Istituzioni politiche e studi di diritto fra Cinque e Seicento. Il 'Messanense studium generale 'tra politica gesuitica e istanze egemoniche cittadine* (Milan: Giuffré, 1994), part. 1, chap. 2.

38. For example, Francesco Sforza turned to rhetoricians at the University of Pavia. Agostino Sottili, "L'università di Pavia nella politica culturale sforzesca," *Gli Sforza a*

Milano e in Lombardia e i loro rapporti con gli stati italiani ed europei (Milan: Cisalpino-Goliardica, 1982), 557-60.

39. In general, Carlo Cipolla, *Public Health and the Medical Profession in the Renaissance* (Cambridge: Cambridge University Press, 1976); Edgardo Morpurgo, "Lo Studio di Padova, le epidemie, e i contagi durante il governo della repubblica veneta," *Memorie e documenti per la storia dell'Università di Padova* 1 (1922): 159; Elia Lombardini, *Dell'Origine e progresso della scienza idraulica nel Milanese ed in altre parti d'Italia* (Milan: D. Salvi, 1860).

40. Jean Delumeau, *Vie économique et sociale de Rome dans la seconde moitié du XVI siècle*, 2 vols. (Paris: De Boccard, 1957-59), 1: 338-39; 2: 580-81. Benedetto Castelli's *Della misura delle acque correnti*, written while on a visiting professorship at the University of Rome in connection with the acqueduct project and providing an exegesis of Frontinus' *De aquaeductu*, is reprinted in *Scienziati del Seicento*, 181-212.

41. Material concerning the projects commissioned by the Venetian government is scattered throughout ASV, *Riformatori*, in particular, filza 75-78; 179-87. The role of Galileo and his disciples is mentioned in Manlio Pastore Stocchi, "Il periodo veneto di Galileo Galilei," *Storia della cultura veneta*, vol. 4: *Il Seicento*, 2: 37-66, Adriano Carugo "L'insegnamento della matematica all'Università di Padova prima e dopo Galileo," *Ibid.*, 151-99, and, indispensably, Antonio Favaro, "I successori di G. Galileo nello Studio di Padova," *Archivio veneto*, n.s. 3, vols. 33-34 (1917).

42. Danilo Marrara, *Riseduti e nobiltà: Profilo storico-istituzionale di un'oligarchia* (Pisa: Pacini, 1976), 131. Difficulties encountered by the government in Bologna are recorded by Celestino Piana, *L'Università di Bologna e il Collegio di Spagna. Nuovi documenti*, 2 vols. (Bologna: Publicaciones del Real Colegio de España, 1976), 1: 144. Compare R. Steven Turner, "University Reformers and Professional Scholarship in Germany, 1760-1806," in Lawrence Stone, ed., *The University in Society*, 2 vols. (Princeton: Princeton University Press, 1974), 2: 495-532.

43. Pietro Accolti's *Parere per riformare lo Studio di Pisa* (1611) is quoted in Giuliana Volpi, "Lineamenti per uno studio sull'Università di Pisa nel XVII secolo," *Scritti in onore di Dante Gaeta*, Pubblicazioni della facoltà di giurisprudenza della università di Pisa (Milan: Giuffré, 1984), 670.

44. An example among many is in ASF, *Archivio Mediceo del Principato*, filza 7, fol. 91r, 30 April 1546 [MAP document 4097], a letter from Cosimo I to Nicolas Perrenot de Granvelle, requesting the imperial minister's assistance in obtaining the services of Andrea Alciati for the Studio Pubblico in Pisa since Alciati's contract with the university in Ferrara has just expired. In order to avoid suspicion of stealing Alciati's services from the Duke of Ferrara, Ercole II d'Este, Cosimo I wished to have the jurist recalled to Milan for service at the university of Pavia and subsequently to have him ordered to Pisa. In addition, for what follows, Antonio Ioli, "Sul soggiorno Messinese di Marcello Malpighi, 1662-6," *Annali di storia delle università italiane* 2 (1998): 157-64.

45. ASV, *Riformatori* filza 81, 5 January 1714.

46. ASV, *Riformatori* filza 191, 29 August 1711.

47. Maria Laura Soppelsa, *Genesi del metodo galileiano e tramonto dell'aristotelismo nella scuola di Padova* (Padua: Antenore, 1974), 117.

48. Pietro Cossali, *Elogio di Giovanni Poleni* (Venice: 1813), 16.

49. ASV, *Riformatori* filza 192, 14 December 1712: "Il a aussi beaucoup d'habilité dans l'Architecture des Eaux, qu'il a apprise en Hollande de sorte qu'il pouvroit servir utilement le Publique dans cette partie si importante."

50. ASV, *Riformatori* filza 195, 3 June 1718: "Non ho lasciato però nel frattempo di tener linguaggio del proposito col Sig. Cavaliere Newton, soggetto di molto accreditata virtù e presidente di questa Società Reale, il quale in primo luogo mi disse esservi in Londra scarsezza di persone intelligenti nelle Bottaniche, e che non me ne potrebbe con-

tare che due o tre, cioè un medico assai erudito di famiglia Sloone [sic, for Sloane] e due speciali. Mi aggiunse poi, bene sapere essere in Italia particolarmente coltivato, e professato tale studio."

51. Michelangelo Fardella, *Lettere ad Antonio Magliabechi,* ed. Salvatore Femiano (Cassino: Garigliano, 1978), 101.

52. Howard B. Adelmann, ed., *Correspondence of Marcello Malpighi,* 4: 1478, to Marcantonio Borghese, 13 July 1689. The average number of students is from Anton Felice Marsili, "Memoria per riparare i pregiudizi dell'Universita degli Studi di Bologna e ridurla ad una facile e perfetta riforma" (1689) edited by Ettore Bortolotti in *Memorie intorno a Luigi Ferdinando Marsili* (Bologna: L'Istituto, 1930), 383-471.

53. Stefano De Rosa, "La riforma e il paradosso," quoting Girolamo da Sommaja, 105.

54. Nicola Carranza, *Monsignor Gaspare Cerati, provveditore dell'Università di Pisa nel Settecento delle riforme* (Pisa: Pacini, 1974), quoting Cerati, 251.

55. Magalotti to Prince Leopoldo, 13 December 1667, quoted in William E. Knowles Middleton, "Some unpublished correspondence of Lorenzo Magalotti in 1667-1678," *Studi secenteschi* 20 (1979): 186.

56. ASV, *Riformatori,* b. 172, dated 28 Febuary 1665. The voting is recorded throughout the seventeenth century. For example, ASV, *Riformatori,* filza 182, date 20 May 1684, on the choice of Leale Leali, second-chair reader in surgery in 1683, Agostino Pivati, reader in the same subject from 1681, and Angelo Montagnana, who became third-chair extraordinary reader in practical medicine in 1687.

57. Anthony Grafton and Lisa Jardine record the notes of the student who commented, "my teacher spoke too fast." *From Humanism to the Humanities,* 87.

58. Richard L. Kagan, "Universities in Italy," Dominique Julia, Jacques Revel, Roger Chartier, eds., *Les Universités européennes du XVIe au XVIIIe siècle: histoire sociale des populations étudiantes* (Paris: Éditions de l'École des hautes etudes en sciences sociales, 1986-1989), 1: 158-59, presents figures for Ferrara and Padua and some fragmentary information on Bologna and Pisa. More figures for Pisa have now been put together by G. Volpi, "Lineamenti per uno studio," 765-83.

59. Paola Zambelli, "Uno, due, tre, mille Menocchio?" *Archivio storico italiano* 137 (1979): 51-90 is a reinterpretation of the data supplied by Carlo Ginzburg, *The Cheese and the Worms,* trans. Anne and John Tedeschi (Baltimore, Md.: Johns Hopkins University Press, 1980, orig. ed. Turin, 1976).

60. The most famous case in the mid-seventeenth century was the quarrel between Alessandro Marchetti and Giovanni Alfonso Borelli. Similar quarrels were influential in the dissolution of the Accademia del Cimento, says Paolo Galluzzi, "L'Accademia del Cimento: 'Gusti' del principe, filosofia e ideologia dell'esperimento," *Quaderni storici* 16 no. 3 (1981): 809.

61. Among the most adept in this approach was Giovanni Battista Riccioli, who filled his *Almagestum novum* (1651) with such "experiments." A few are mentioned in Paolo Galluzzi, "Galileo contro Copernico: il dibattito sulla prova 'galileiana' di G. B. Riccioli contro il moto della Terra alla luce di nuovi documenti," *Annali dell'Istituto e Museo di Storia della Scienza* 2, no. 2 (1977): 87-148.

62. Paolo Galluzzi tries to turn the event into another 1632 in "Libertà scientifica, educazione e ragione di stato in una polemica universitaria pisana del 1670," *Atti del 24 Congresso Nazionale di Filosofia, L'Aquila, 28 aprile-2 maggio 1973* 2 (Rome: 1974): 404-12; Danilo Marrara, "Le cattedre ed i programmi d'insegnamento dello Studio di Pisa nell'ultima età Medicea (1712-37)," *Bollettino storico pisano* 51 (1982): 124, places it in a more moderate perspective. The authority on the atheist trials is still Luciano Osbat, *L'Inquisizione a Napoli: il processo agli ateisti 1688-1697* (Rome: Edizioni di Storia e Letteratura, 1974).

Nature and the Universities 85

63. This was the tactic of Alessandro Marchetti, quoted in Danilo Marrara, "Le cattedre ed i programmi d'insegnamento," 121.

64. ASV, Riformatori, b. 430, n.d., unnumbered fols.

65. Giuseppe Ricuperati, "L'Università di Torino e le polemiche contro i professori in una relazione di parse curialista del 1731," *Bollettino storico-bibliografico subalpino* 64 (1966): 366.

66. Marta Cavazza, *Settecento inquieto* (Bologna: Il Mulino, 1990), chap. 1; Salvatore Rotta, "Scienza e 'pubblica felicità' in Geminiano Montanari," in *Miscellanea Seicento* Università di Genova, Istituto di Filosofia della Facoltà di Lettere e Filosofia (Florence: Le Monnier, 1971), 95; Michele Maylender, *Storia delle accademie d'Italia*, 5 vols (Bologna: Cappelli, 1925-9), 4: 34. Concerning the Fisiocritici, Chiara Crisciani, et al., *Scienziati a Siena* (Siena: Fisiocritici, 1999); concerning the Investiganti, Max Fisch, "The Academy of the Investigators," *Science, Medicine and History. Essays on the Evolution of Scientific Thought and Medical Practice Written in Honor of Charles Singer* ed. E. Ashworth Underwood (Oxford: Oxford University Press, 1953), 521-62; Maurizio Torrini, "L'Accademia degli Investiganti, Napoli, 1633-70," *QS* 16 (1981): 845-83. Concerning il Cimento, Paolo Galluzzi, "L'Accademia del Cimento."

67. Marsili's text is reprinted in Ettore Bortolotti, "La fondazione dell'Istituto e la riforma dello Studio di Bologna," *Memorie intorno a Luigi FerdinandoMarsili* (Bologna: Zanichelli, 1930), 383-471. An appreciation of Marsili as a natural philosopher is in Gregorio Piaia, *I filosofi e le chiocciole. Operette di Anton Felice Marsili, 1649-1710* (Assisi: Porziuncola, 1995). In addition, Marta Cavazza, "Riforme dell'università e nuove accademie nella politica culturale dell'arcidiacono Marsili," in *Università, accademie e società scientifiche in Italia e in Germania dal Cinquecento al Settecento*, ed. Laetitia Boehm and Ezio Raimondi (Bologna: Il Mulino, 1981), 245-82, now in Cavazza, *Settecento inquieto*, chap. 2.

68. Biagio Brugi, "Un parere di Scipione Maffei intorno allo Studio di Padova sui principi del Settecento," *Atti dell'Istituto Veneto di Scienze, Lettere ed Arti* 69 (1909-10): 588; Idem, "Scipione Maffei e lo Studio di Padova," *Atti dell'Istituto Veneto di Scienze, Lettere ed Arti* 68 (1908-9):P 897-903; Maria Laura Soppelsa, "Itinerari epistemici e riforme istituzionali nello Studio di Padova tra Sei e Settecento," Luigi Olivieri, ed., *Aristotelismo veneto e scienza moderna. Atti del XXVmo Anno Accademico del Centro per la Storia della Tradizione Aristotelica nel Veneto* 2 vols. (Padua: Antenore, 1983), 2: 961-92; Giuseppe Ricuperati, "L'università di Torino nel Settecento," *QS* 8 (1973): 575-98; Gian Paolo Romagnani, "Scipione Maffei e il Piemonte," *Bollettino storico-bibliografico subalpino* 84 (1986): 133-227.

69. The one the Tuscan government established at Pisa was exemplary. Augusto Occhialini, *Notizie sull'Istituto di Fisica Sperimentale dello Studio di Pisa* (Pisa: Mariotti, 1914), 3-4; also, Gian Antonio Salandin and Maria Pancino, *Il 'teatro' di fisica sperimentale di Giovanni Poleni* (Trieste: Lint, 1987); Giuseppe Ermini, *Storia dell'Università di Perugia*, 2nd ed., 2 vols. (Florence: Olschki, 1971), 1: 574.

70. I discuss University reform at Padua in my "Giornalismo, accademie e organizzazione della scienza: tentativi di formare un'accademia scientifica veneta all'inizio del Settecento," *Archivio veneto*, ser. 5, vol. 120 (1983): 5-39.

71. Giulio Guderzo, "La riforma dell'Università di Pavia," *Economia, istituzioni, cultura in Lombardia nell'età di Maria Teresa*, 4 vols., ed. Aldo De Maddalena, Ettore Rotelli, Gennaro Barbarisi (Bologna: Il Mulino, 1982) 3: 850, citing Paolo Frisi.

72. Nicola Carranza, *Monsignore Gaspare Cerati, provveditore dell'Università di Pisa*, 239.

73. Franco Venturi *Settecento Riformatore*, vol. 2: *La chiesa e la repubblica dentro i loro limiti (1758-1774)* (Turin: Einaudi, 1976), 180; Eric Cochrane, *Florence in the Forgotten Centuries, 1527-1800. A History of Florence and the Florentines in the Age of the*

Grand Dukes (Chicago: University of Chicago Press, 1973), 450-91; Cesare Mozzarelli, *Sovrano, società e amministrazione nella Lombardia teresiana, 1749-58* (Bologna: Il Mulino, 1982), 48-56, 185-213; Christof Dipper, *Politischer Reformismus und begrifflicher Wandel. Eine Untersuchung des historische-politischen Wortschätzes der Mailander Aufklärung, 1764-96* (Tubingen: Niemeyer, 1976), 13-29.

74. Baldo Peroni, "La riforma dell'università di Pavia nel Settecento," in *Contributi alla storia dell'università di Pavia* (Pavia: Università, 1925), 149.

75. The best account of these reforms is still Piero Del Negro, "L'Università," *Storia della cultura veneta*, vol. 5: *Il Settecento*, 2 parts (Vicenza: Neri Pozza, 1985), 2: 47-76. In addition, Alba Veggetti and Bruno Cozzi, *La scuola di medicina veterinaria dell'università di Padova* (Trieste: Lint, 1996); as well as Giuseppe Armocida and Bruno Cozzi, *La medicina degli animali a Milano: i duecento anni di vita della Scuola veterinaria (1791-1991)* (Milan: Sipiel stampa, 1992).

76. A few of these reforms are analyzed by Elena Brambilla, "Tra teoria e pratica: studi scientifici e professioni mediche nella Lombardia settecentesca," *Lazzaro Spallanzani e la biologia del Settecento* (Florence: Olschki, 1982), 553-68.

77. Eluggero Pii, "Le origini dell'economia 'civile' in Antonio Genovesi," *Il pensiero politico* 12 (1979): 330-34; Ugo Marcelli, "Insegnamento e pratica delle dottrine economiche a Bologna nel secolo diciottesimo. Giacomo Pistorini, lettore dello Studio e consultore del Senato," *Studi e memorie per la storia dello Studio di Bologna*, n.s., 1 (1956): 487-503. Compare Andrea Gardi, "L'università di Ferrara come terreno di scontro politico-sociale all'epoca di Benedetto XIV," *'In supreme dignitatis,'* 309-38.

78. Piero Del Negro, "I 'Pensieri di Simone Stratico sull'Università di Padova' (1760)," *Quaderni per la Storia dell'Università di Padova* 17 (1984): 32.

79. Quoted by Franco Venturi in "Postilla," *Annali della fondazione Luigi Einaudi* 19 (1985): 454. Here I follow the interpretation of Ferrone, *I profeti dell'Illuminismo. Le metamorfosi della ragione nel tardo Settecento italiano* (Bari: Laterza, 1989), 57-60.

80. Mario Saibante, C. Vivarini and G. Voghera, "Gli studenti all'Università di Padova dalla fine del Cinquecento ai nostri giorni," in M. Saibante, *Studi di economia, statistica e sociologia, 1924-55*, ed. Corrado Gini (Rome: Rivista di politica economica, 1959), 715-77.

Chapter 5:
Teaching and Learning

The notoriously varied demands on students' attention outside the classroom in modern universities are nothing new. In a report to the governing board of the University of Padua, one early eighteenth-century law professor noted:

> They come to the university with the certainty of having to remain for four years. With this in mind, they just barely get to know the professors enough by sight to greet them on the street, while they do anything except those things which are conducive to study; and having formed this habit in the first year, they continue fatally till the last year of their degree, in which, overwhelmed like the foolish virgins of the gospel, without oil in their lamps, they run desperately to the professors with the inopportune, "date nobis de oleo vestro."[1]

Extracurricular activities included gaming, whoring, drinking and dueling as well as offers of menial positions in the houses of the great—the early modern equivalent of internships. Yet, who could blame the students? Even the more diligent soon discovered that materials taught in class often had little to do with what they needed to know for life—much less for passing exams. Small wonder that they searched for other diversions, or even other ways of becoming educated. They joined the number of matriculants, and eventually of degree recipients, who filled the statistical tables without filling the classrooms. A worried university chancellor Bortolo Sellari at Padua in 1723 reported that lectures were deserted except for "a few Greeks, [and] the others are nothing but the friends of the Public Professors."[2]

To be sure, techniques for engaging the attention of listeners existed long before attendance certificates were finally instituted as a last resort. Already in the early seventeenth century Federico Cesi bitterly denounced professors who abased themselves by using rhetorical pyrotechnics, "giving pleasure and bringing forth high-sounding and sonorous doctrines, careless of whether the opinions are true as long as they are plausible, magisterial and authorized by the more common views of the ruling sect."[3] In Cesi's view, however successful such techniques might be in attracting students, they were not likely to redound to the benefit of science over the long term.

How could the interests of students and of science be reconciled? We will examine some cases in this chapter. Toward the late seventeenth and early eighteenth centuries, professors tried new teaching techniques, with some success. By incorporating their own research into their teaching in novel ways, they sought to appeal to the public without sacrificing rigor. And in the careers of Giovanni Girolamo Sbaraglia, Antonio Vallisneri and Giovanni Poleni at the Universities of Bologna and Padua, we will see how the modern profession of university professor was born.

What modern college science teacher has never voiced sentiments analogous to those of Antonio Vallisneri, at the beginning of a summer recess? "I've waited patiently to be relieved of the immense weight of my public lessons without enjoying myself by researching the most curious and most recondite operations of nature . . . now I return to look again at my old studies, ever widening their compass according to Nature's greatness and wondrousness."[4] In Vallisneri's time, a great divide had already begun to open up between the two functions of teaching and research. To turn the two halves of the college profession into a harmonious whole required more than the ingenuity of a few professors could accomplish.

The ambiguous attitudes of the officials in charge of universities were partly to blame for the dichotomy between teaching duties and research interests. Officials favored traditional styles of teaching because of the need to prepare students for entry exams into the professions. Yet they behaved at times as though education could serve useful political purposes if only it would begin to reflect some of the most recent intellectual trends.[5] But after they founded the first chairs of botany, humanities, and anatomy and planted the first gardens of simples, they did very little else before the eighteenth century to make sure university teaching would develop and change with the changing knowledge of the professors who imparted it—except by their hiring policies.

Small wonder that the teaching of many professors became almost entirely detached from the moorings of the statutes. And while Galileo and, as we suppose, Malpighi, at least made a pretense of commenting on the accepted authors,[6] Claude Bérigard tried to introduce a whole new set—the Presocratics. At Pisa from 1627-39 and then at Padua until his death in 1663, he adopted this approach as the best way to introduce Galileo's critique of qualitative physics, and eventually, Descartes' notion of particles and motion as the basis of all matter and William Harvey's discovery of the circulation of the blood.[7] So out went the old Aristotelian texts upon which disputation questions were usually based and in came a new series of ancients of unimpeachable antiquity. Had not Aristotle only criticized them out of "levitas et invidia"?[8] Yet "there is nothing else we can do," he said, to discover the ideas of these authors "than to imagine what they might have thought," since Aristotle's account of them, the only one we have, is full of distortions."[9] In this original fashion, Bérigard left open the blanks to be filled by modern ideas.

Geminiano Montanari, at Bologna and Padua in the 1680s, used the difficulties raised in recent research concerning Aristotle's *Meteorologia* and *De coelo* as an occasion to expand his astronomy course into a general discussion of the principles of science.[10] He cautioned against uncritical acceptance of Descartes' account of the four elementary types of particles and their power to generate phenomena by their motion. He even rejected Galileo's cautious assertion that some indivisible point-like atoms without extension ("parti non quante") must exist.[11] "It is not philosophical to determine what precise shape these particles might have," he noted, "if we do not have other conjectures to back up our argument."[12] He praised Boyle for not having gone beyond those few conclusions that could safely be drawn based on the empirical data against the four elements of the Aristotelians and the three principles of the chemical philosophers. That method, he said, was exactly the opposite of the method of Aristotle, who "attached himself from the beginning to those conclusions that he should have ended up with."[13]

These Italian professors thus joined the many of their colleagues in other parts of Europe who presented science not as an unchanging set of doctrines but as a method of discovering truth by eliminating improbable conjecture. Likewise, Christian Wurstisen taught the Copernican system at the University of Basel, while Jean Du Hamel at the University of Paris and Jean Chouet at the College of Geneva tried to give Cartesianism a fair hearing.[14] So seriously, indeed, did the professors take their educational duties that some of them appear to have seen little difference between the task of preparing lessons and that of preparing scientific tracts. Those fortunate enough to teach in their fields of research, both in Italy and abroad, therefore polished up their lessons from time to time for publication.[15]

Were the efforts of these professors always and everywhere appreciated? At least to some extent, student reactions may have depended on how helpful their lessons could be for passing exams. In Montanari's case, students voted with their feet. Of the twenty students who began his course, "there are no more than two who pass the second book of Aristotle's *Meteorology* without leaving me."[16] As long as other professors at the university continued to teach the standard texts in the traditional manner, Montanari's ex-students had somewhere to go—that is, if they did not simply stay at home and hire private tutors—and university enrollments were safe. To these students it scarcely mattered that the professors who imparted instruction in the material required for exams were often precisely those whose involvement in pan-European scientific debates was less pronounced.

A few professors managed to combine education—both of the public at large and of a few advanced students—with voluminous scientific publication in an effort to expand the traditional limits upon the definition of their profession. The new methods of teaching were more frequently found among the opponents of the standard traditional curriculum; but not exclusively. By using them, traditionalists may well have contributed unwittingly to the advancement of the innovators. In this connection, let us now examine the interesting case of Giovanni Girolamo Sbaraglia at the University of Bologna. A self-proclaimed "Galenist," Sbaraglia resorted to in-class demonstrations to prove the verity of the doctrines of his chosen intellectual model—as well as, at times, to criticize them.[17] His student Antonio Vallisneri, one of the great innovators of late seventeenth-century biology, carefully followed his example.

The basic message of Sbaraglia's lessons, at least in the time that Antonio Vallisneri took notes from him around 1684, was about the importance of using experiment and observation to criticize various opinions.[18] For instance, in his lessons on the upper body, he called attention to Hippocrates, who claimed that teeth had the quality of preferring heat to cold—a theory supported, he said, by the "communis opinio" confusing the nerve with the tooth itself in spite of the difference in composition. He told of an experiment in which the tooth was deeply penetrated by a pointed instrument without causing any sensation of pain in the (perhaps excessively stoic!) patient.[19] And he did not limit himself to recounting these experiments; but he involved the students in carrying them out. In a lesson on the circulation of the blood based on the theories of William Harvey, he had them pull up their sleeves to expose an artery. "Press it with your fingers," he told them, "at each pulsation." As the blood passed through the artery on its way to the extremities, the pulsations seemed to change according to the amount of blood allowed to pass.[20] Therefore, he argued, the pulsations depended on the pressure of the blood.

Sbaraglia demonstrated also how experiments could be interpreted wrongly. For instance, he described what he called an experiment by Galen, repeated in the sixteenth century by Girolamo Fracastoro, concerning the pulsations of the blood. In this experiment, certain hollow pieces of straw were slipped inside an artery that had been cut for this purpose, after which the artery was tied beneath the cut. The flow of blood from the opening of the straw, according to the experimenters, showed that the arteries contained blood and not air, whereas the absence of pulsations in the blood flowing out, in spite of the obvious pulsing motion of the walls of the artery above the ligature, showed that the pulsation was an inherent faculty of the arteries or else was transmitted to them by the heart by way of the artery walls. "But I tried this experiment," Sbaraglia told his students, "and it proves nothing." The absence of pulsations in the blood coming from the straws, he explained, was due to the motion lost crossing the narrow opening of the straws.[21]

However, Sbaraglia made no secret of his disinterest in new investigative methods for any other purpose besides perfecting therapeutic techniques. "Philosophizing is not the same as curing," he liked to say. In Sbaraglia's view, rather than wasting their time with such inanities, students ought to dedicate themselves to working with patients. He condemned professors like Malpighi who (according to his somewhat exaggerated view) did not bother to cure the sick; and instead, he sought to emulate his own maestro, Giovanni Antonio Cucchi, who had divided his time between his university chair at Bologna and his medical practice in Comacchio.[22] As far as basic medical knowledge was concerned, the Ancients were still good enough. Even the invention of the microscope had only in a very few cases modified the basic conceptions of Galen concerning the use of the parts of the body and the way of effective curing.[23]

According to Sbaraglia, familiarizing oneself with the medical ideas of the Ancients was far more important than adding to them. In his lessons on physiology, he gave a literal and traditional interpretation of Galen's doctrine on the three spirits. During the formation of the fetus, he explained, the animal, vital, and natural spirits maintained the vital heat and humidity originally produced by the combination of the male and female semen. Next, they took their respective places in the principal parts of the body, brain, heart, and liver. These spirits, he affirmed, caused all sensations and movements in the mature organism. The four humors, on the other hand, blood, phlegm, bile, and black bile, circulated in various quantities to determine the states and temperaments of the body. Health, according to this hypothesis, consisted in maintaining equilibrium between the four qualities, i.e., hot, cold, humid, and dry, that manifested themselves in the humors. Sbaraglia's presentation therefore was not much different from that of Galenic treatises by the likes of Lazare Rivière, from whom he adopted entire arguments.[24]

Sbaraglia defended the system of the Ancients against the mechanical conception of physiology championed by Descartes, Giovanni Alfonso Borelli, and Malpighi, based on the movement of the particles of matter.[25] Harvey's theory of the circulation of the blood, he insisted, was no better than the ancient ideas. For all the novelty of Harvey's concept of a mechanical role for the heart in the motion of the blood, did he not nonetheless continue to view the heart as the center of the microcosm and of the vital force?[26] Furthermore, Sbaraglia insisted that Jean Baptiste van Helmont's pathology based on disturbances in the "archaeus" or life force within each living thing was far inferior to the ancient physiological theories of the four qualities, the four humors, and the temperaments based on these.[27]

Using logic and reason, Sbaraglia tried to introduce tiny changes into the ancient system. The doctrine of the three spirits, he admitted, with Santorio Santorio, contained serious contradictions. The natural spirit could not possibly originate in the liver, which was obviously built only for filtering excrements from the blood and nothing else, as recent research by Malpighi and Lorenzo Bellini had shown. Fortunately, the physiology of Galen worked just as well with the two remaining spirits, vital and animal, in spite of the negative opinions of contemporary authors like Rivière, who preferred a more literal reading.[28] But the elimination of the natural spirit created other problems and necessitated other changes, since the traditional texts traced the vital spirit to a combination of natural spirit with air from the lungs. The work of the Spanish theologian and physician Michael Servetus offered a way out: vital spirit was constituted from an ethereal part of the blood through the action of the heart and lungs.[29] Finally, Sbaraglia sought to modify the metaphysical content of the traditional doctrine, suggesting that the role of the spirits was not explicable only in terms of the divine nature attributed to them by Hippocrates, but also in terms of mechanical principles. "Non est opus deos in scenam advocare," he said, citing Harvey.[30]

At first, Vallisneri was highly influenced, even in his method of curing, by these lessons of Sbaraglia—whose "erudition" he "admired."[31] Later, however, after taking his degree in Reggio Emilia and visiting the greatest physicians and pharmacists of his time (Pompeo Sacco in Parma, Lelio Trionfetti in Bologna, and, in Venice, Bernardo Florio, Jacopo Grandi, and Lodovico Testi), he abandoned the idea of dedicating himself entirely to medicine and began his first important studies on insects. This new interest in scientific research directed him to a university career. With the publication of his first results in the *Galleria di Minerva* printed in Venice (1694 and 1700), the Riformatori dello Studio di Padova in 1700 gave him a chance to realize his ambition. On the recommendation of the recently hired Sacco, they offered him the first chair of practical medicine at the university. Faced with the prospect of preparing some forty lessons on material that, according to the triennial program, he would be repeating only once every four years, Vallisneri began to take a new look at the notes he had taken from Sbaraglia years before.

On the same sheets where he had taken his notes from Sbaraglia, Vallisneri began to scribble his own personal ideas, inserting his latest discoveries in ever more effective ways. In the part dedicated to *de pulsibus et urinis*, i.e., the circulatory system, he occasionally adopted entirely the text of Sbaraglia, correcting a few errors and refining some of the less elegant expressions.[32] Mild convulsions, he specified, were caused not by general inflammation of the heart, but by that of the ventricles, the nerves, and the muscular fibers. He now explained Galen's errors by an elegant account of their probable origin.[33] But he showed much more willingness to attack the traditional systems than had Sbaraglia. The progress of knowledge, he insisted, depended on the willingness of researchers to seek new solutions. "[The idea of the] pulsating faculty," for instance, he said, "was imposed for ages upon students." Abandoning it forever called for freeing oneself from the "tyrannical yoke" imposed by the method of learning medicine "only from commentaries."[34] Even more radical were his changes in teaching method. He shortened the part of his exposition devoted to discussing the various contrasting ancient theories and added new experiments concerning the chemical composition of the urine.[35] "The ancient fathers deserve to be praised," he explained, "but experiments, and the truth as it is drawn out of obscurity, deserve to be praised even more."[36]

The lessons on practical medicine gave Vallisneri his first chance to put these programmatic statements into action. He began, as the tradition specified, by presenting the problem drawn from the ancient text.[37] Since his teaching was supposed to include all the diseases of the head, thorax, and lower abdomen, the text in question was supposed to be the standard guide to Galenic physiology as a whole: namely, Avicenna's *Liber Canonis*, in particular the first introductory book, the third book on diseases of the whole body, and the first *fen* of Book 4 concerning fevers. In his series of lessons on fevers, Vallisneri introduced Avicenna's doctrine based on the notion of the heart as the distributor of so-called radical heat and humidity. "Fever," he affirmed, citing the text, "is an extraneous heat that ignites in the heart and proceeds from this through the whole body by way of the spirit and the blood in the arteries and veins."[38] Similarly, he began the series of lessons on diseases of the head with Avicenna's doctrine on headaches.[39] "[Headache] is an injury that afflicts the head. Now, all injuries are caused by a change or by a conflict of temperaments," Avicenna had said, referring to the notion of the balance of *complexio* or temperament, i.e., of the four qualities (hot, cold, humid, dry) characteristic of the healthy body—"or else from a contusion, or from both of these causes."[40]

Instead of continuing with an exposition of the ancient text, Vallisneri proceeded to give a detailed account of his own observations. In that period, however, most of his work concerned insects, and therefore had little to do with the material at hand. But this did not stop him. In the course of his studies on worms existing within the human body, to be published in 1703, he had deduced that some headaches were caused by the movement of worms and the deposit of eggs in the brain and nose. On this topic, to which Avicenna had alluded in a single paragraph, he dwelt for no fewer than four lessons. He spoke so extensively concerning his observations on worms that after the first year he began to worry about dragging students through this material for his own personal reasons. "I did not do these lessons later on," he said in a personal note, "in order not to bore students by a surfeit of worms."[41]

Rather than cautiously defending the Ancients, Vallisneri showed how the observations of modern scholars had radically changed general ideas about medicine. True, he noted, some researchers had drawn excessively daring conclusions, as in the case of the English physiologist Thomas Willis, who, considering the properties of sulfur and of the blood, had erroneously suggested that headaches could be caused by excessive sulfur brought into the brain by the blood—a conclusion that later research had shown to be unfounded. "We are not so dependent upon the latest authors, nor so independent [from the ancient ones]," said Vallisneri, "that we have to subscribe blindly to all their opinions."[42] In general, however, he noted, the new discoveries had contributed a deeper knowledge of the functioning of the body. His contemporary, Lorenzo Bellini, he added, had shown that the true source of natural heat was motion and not the heart per se.[43] He concluded by urging his students to consult the journals, the academic proceedings, and the various collections of papers in order to stay up to date on the latest developments in research.

In ten years of constant teaching according to this method, Vallisneri contributed to the gradual change that was going on in the aim and function of university lessons. Called in 1710 to teach theoretical medicine to students in their second year, he entirely abandoned the ancient text. Instead, he introduced the approach of Galileo's philosophical treatises—that is, of referring to ancient writers only as brilliant historical precedents, methodological models, and rhetorical examples. He omitted Galen entirely from the title of his course and proposed to speak "On Reproduction." In fact, Galen's *Ars Medica*, the text on which he was supposed to comment, provided a synopsis of medical philosophy in general, but nothing at all on reproduction. Setting aside entirely the chapters in Galen, he organized his course any way he wished. Since Galen furnished him with nothing other than an ancient authority to correct in the light of recent discoveries, he took considerable liberties in his citations. For instance, in the definition of the function of the testicles, he stitched together phrases from two different works of Galen—the *Ars Medica* and the *De Semine*.[44] And once again, according to his own research interests, soon after the section "On the Ovaries" he inserted one "On Worms."

Using his own microscopic observations, Vallisneri corrected the theories of the ancient naturalists as well as his transalpine and Italian contemporaries and presented the latest results to his students. In the first version of these lessons he rejected all the most venerable ancient theories—Galen's notion about the mixing in the uterus of the potent male semen and the weak female semen, Aristotle's notion of the menstrual blood as the prime material on which the male semen operated. He explored a concept that had been mentioned but not developed by the Ancients, suggested that the male and the female both had an active role in reproduction; and he reduced reproduction to a simple mechanical action. He refuted the Galenic theory about the possibilities of spontaneous generation, verifying his conclusions by detailed diagrams of the hitherto unknown reproductive apparatus of the smallest animals. Then he refuted Antonie van Leeuwenhoek's preformationist theory, too much based, in his view, on a notion of female passivity, and he presented to students examples of erroneous observations made by defective instruments, such as might have suggested such a theory to Leeuwenhoek.[45] He rejected the traditional idea of the ovaries as a sort of female testicles and, in accord with Regnier De Graff, described them instead as eggs of a sort, equal in importance with the semen. In fact, he believed that the microscopic spermatozoa had no other purpose but to excite the female organs in such a fashion as to make fertilization possible, and saw the moment of conception in terms of the reawakening of tissues by the stimulating effect of the seminal liquid. But he did not desire that students should go away thinking they had heard the ultimate answer to the problem of reproduction; so he gave them a humanist parable: "The disciples of Pythagoras," he began, "so venerated the oracles of their maestro, that when anyone asked why they reasoned in a certain way, they always responded, because he himself, i.e., Pythagoras, had said it." He hoped his students would be able to say they had gained much more than mere information, which in any case would eventually be replaced by other better information. "I do not want you to be tied down . . . similarly by the authority of your maestro," he told them. "I want you to consider my reasoning only in order to say, not he himself has said it, but Nature has said it."[46] This, he believed, was the true method of the ancient natural philosophers, and the most useful for the progress of knowledge.

This new teaching method, used by Vallisneri and others at Padua and Bologna and elsewhere, had some disadvantages. The father of a young student insisted "in the confusion of the sciences and the competition of the curious minds one finds unhealthful and excessively novel beliefs."[47] Yet there was no proof that the students of a Vallisneri, with all their new ideas, were better prepared, from the point of view of public health, than those of a traditionalist like Sbaraglia. Some of the therapeutic techniques of Vallisneri were, in spite of the highsounding theories, still rooted in ancient practices. Against worms in the brain, he suggested modest quantities of oil. To stimulate the intestine, he recommended cassia mixed with cream of tartar, and, in the most difficult cases, a potion based on cassia, malva, and powdered crab's eye, taken five hours after lunch. To move the stagnant fluids in the body, he suggested the cauterization of the muscles at the temples. And when nothing else served to alleviate the pressure of the fluids in the brain, especially in the case of venereal disease, he prescribed an ancient but effective remedy: the trepanning of the cranium.[48] Between these remedies and the Galenists' favorite technique of bloodletting, what were the university officials to choose? Instead of choosing, they encouraged both.

In the insistence on public utility, as well as on public education, Vallisneri was matched, and even surpassed, by Giovanni Poleni. A professor at Padua from 1709 to his death in 1762, Poleni is now almost exclusively remembered for his contributions to mechanics and its applications. Besides his role in the development of hydrostatics, kinetics, the stress principles of structural design, he contributed to the editing of scientific texts.[49] In his own century he was also remembered as a great teacher, whose most brilliant student, Vitalino Donati, attained "international importance" in a professorship at Turin.[50] During his own lifetime he was regarded by the Venetian government's officials as their precious consultant on all engineering questions dealing with the lagoon. By association with them he acquired their habit of saving every scrap of paper that had anything to do with official duties. Since he conceived of his profession of university professor as constituting an official duty resembling his others, he became a far more careful archivist of his personal papers than many contemporaries surer than he of having made an immortal contribution to learning. Among the things he kept, which also included thirty-odd volumes of personal correspondence and several unpublished treatises on subjects ranging from optics to moral philosophy, were several fascicles of his daily university lessons.[51] Annotated, apparently, up until the last moment before delivery, full of bibliographical indications and illustrated by sketches for visual aids, this documentation chronicles a long career dedicated to communicating an important message to undergraduates and the public at large.

Poleni did not begin his career at Padua with much personal experience of what university teaching should be like. Yet he was fortunate to have received some of the best alternative education available to Venetian citizens in his period at the Collegio alla Salute maintained by the Clerks Regular of Somaschi under the auspices of the Riformatori dello Studio.[52] And he was still more fortunate to have had as his tutor—with the close personal relationship characteristic of Somaschi pedagogy—one of the most learned members of the order in Venice: Francesco Caro, author of a five-volume compendium of lessons entitled *Philosophia amphisica ex Aristotelis Democratique mente illustrata* (Venice, 1695).[53] Since Caro's ambition was to ensure "that my students should come regularly to school,"[54] by promising them lessons in which their attention was to be rewarded by exciting discoveries, Poleni did not just copy down an exposition of the adulterated Aristotelianism that this work's curious title seemed to announce. He did not even really learn the unusual mixture of modern authors to whose doctrines Caro, later on in the text, claimed chiefly to adhere: "Bacon on logic, Gassendi on physics, Tycho on Astronomy, Galileo on motion, Boyle on experiments, Kircher on mechanics, Cabeo on meteors"[55]). Instead, he learned a complex syncretism. Indeed, his teacher was so determined to put across Gassendi's atomistic message that even the examples for types of syllogism were drawn from chemistry: "every mixed body is made of atoms; every flower is a mixed body; therefore, every flower is made of atoms."[56] Boyle's experiments with pumps and bell jars were apparently performed to prove the existence of some kind of vacuum. Yet the argument against occult qualities drew unexpected support from Descartes' *Dioptrique*. "I will follow whatever reason, experiment, and truth suggest," proclaimed Caro.[57] Poleni's frequent warnings to his students against accepting untested hypothesis and his unwillingness to adhere faithfully to any one system may have been inspired by this education.

Rather than proceeding to the university after graduating from the Somaschi college, Poleni turned to the informal science instruction available among the various religious orders in Venice. Accordingly, he attended the "school" of Tommaso Pio Maffei, a Neapolitan Dominican friar resident at San Giovanni e Paolo in Venice.[58] Maffei, after having been considered for a post in theology at

Padua and dropped as a candidate because of his advanced opinions, had finally been invited to Venice by his Order in the 1690s on the recommendation of certain of the more inquisitive members of the patriciate. Since the Senate looked askance at extra university cultural organizations, he was never encouraged to form a formal scientific academy on the model of the Cimento of Florence, the Investiganti of Naples, or the Inquieti of Bologna.[59] Instead, he held his demonstrations and gave his discourses in the noble household of the Duodo family and in his convent. With "students" like Bernardino Zendrini, later hired by the Senate as an engineer, Bernardo Trevisan, a noble virtuoso, and Antonio Conti, future acquaintance of Newton and Voltaire, his meetings may have produced much useful collaboration. According to Conti's recollection, he repeated many of Giovanni Battista Riccioli's experiments for measuring the speed and force of falling bodies and even invented a simple didactic device involving a pole marked off in degrees from which various weights could be dropped at different heights.[60] To Poleni's education he may have contributed something about the methods for quantifying experimental results.

When Poleni left the protected environment of this private, individual instruction to teach in the university in 1709, he made sure his faltering delivery was amply counterbalanced by his novel exposition. He did not even mention the first book of the *Meteorology* on which he was scheduled to comment. And he left no doubt that "superlunary" and "sublunary" phenomena would be treated as observable data from which the properties of matter and the forces in nature could be deduced rather than as effects of eternal celestial motion that the interaction of the four elements could explain. He praised the wisdom of the Riformatori who had united the fields of astronomy and meteorology under a single professor because of the probability that the study of terrestrial meteorological phenomena could aid in speculations about the atmospheres on the many other possible worlds in the universe—"according to observations, the supposition that every star is surrounded by an atmosphere no different from our earth is shown to be the most reasonable"[61]—and not because of any supposed communication between the spheres. Finally, he showed how the most recent geometrical analysis would be used in his course by assuring his audience that his discussion of comets would follow Newton, "who by the admirable eloquence of his sublime genius so diligently emptied material from the celestial spaces,"[62] and not the popular Cartesian hypothesis. He did not promise a scandal, which was what many members of his audience always expected from the astronomy professor. But he assured them that he would explain the heavens at least as well as his more cautious predecessors—from Andrea Argoli to Stefano Degli Angeli to Montanari[63]— by making the usual political allusion to the condemnation of Galileo and then positing a moving earth ("Let us pretend the earth moves").

In the section on astronomy for the year 1714-15, instead of burdening his students with an unfamiliar technical apparatus, he preferred to introduce them to astronomy by way of those aspects for which the usual texts in ethics, rhetoric, and geography were still sufficient preparation.[64] Not because he was no expert. With one of the world's great telescopes easily accessible at Palazzo Correr in Padua—constructed under Montanari's guidance well before John Flamsteed conceived the Royal Observatory at Greenwich[65]—Poleni had had little trouble in continuing the university's long tradition of original astronomical observations and communicating them from time to time to Daniel Bernoulli.[66] But his research was not well-adapted to the eighteenth-century classroom, where most of his students possessed no more than the rudiments of geometry, where the program was so constructed that he had to make allowances for a large percentage of first-year students every year, and where his only visual aids were a blackboard ("pietra nera," as he called it)[67] and some chalk.

Accordingly, Poleni explained the objects of astronomy as celestial events that occurred through time in a way analogous to the political changes observable in civil history. After a section on chronology, including references to all the modern authorities, from Georg Horn and Isaac Voss to Claude François Dechales and Stanislas Lubienski, he set out to prove the necessity of astronomical knowledge. Unusual celestial events, he argued, seemed to occur in relation to historical ones, like the comets that coincided with the birth of Alexander the Great, the Roman conquest of Egypt, and the death of Pope Innocent VIII.[68] Since the story of man and the story of the heavens were intertwined, an accurate description of the cosmos was of vital importance to all.

Poleni showed how a more accurate astronomy had been made possible by combining the theorems of geometry and arithmetic with the laws of optics.[69] Improvements in scientific instruments occasioned Galileo's discoveries of the phases and apparent changes in size of Venus and Mars in Book 3 of the *Dialogue Concerning the Two Chief World Systems*. In the only visual demonstration in this section, Poleni covered the blackboard with diagrams, drawn from André Tacquet's elementary astronomy manual, on the use of parallax to determine planetary position.[70] Citing observations by Borelli and Cassini on the behavior of comets, he proceeded to the causes for the forces in nature. Dividing his argument by the scholastic quadruple scheme he considered appropriate to metaphysical discourse, he explained that the formal and material (secondary) causes for the generation of the celestial bodies were light, heat, and stellar matter. The efficient and final (primary) causes reposed in God, who "by his infinite wisdom so organized this machine of nature that the secondary causes proceed from the primary causes," and who demonstrated his will to man by the unpredictable celestial events attested by the Psalmist in Psalm 115.[71]

The section on meteorology lent itself to an entirely different teaching methodology. The material Poleni wished to impart was based on observations that did not always require excessively bulky equipment or time periods too long for the class schedule. He had plenty of glass apparatus and pumps on hand, and what he did not have he was in the habit of borrowing from his friend Christian Martinelli, the new custodian of the observatory in Palazzo Correr. Whether or not he actually showed students some of his equipment and, as seems probable, performed some simple experiments in class, the descriptions he gave were so detailed and so well illustrated by diagrams both on the blackboard and in his handouts that students must have felt as though they had been in a real laboratory even when they had never left the lecture hall.[72]

Poleni chose experiments that demonstrated some of the most important scientific discoveries of the seventeenth century. He explained how the Torricelli barometer proved Galileo's error in conceiving the air as a substance with little weight and no pressure.[73] After showing how different lengths of tube could be used to prove that the suspension of the liquid was caused not by a sucking power from within but by external conditions,[74] he suggested other variations to demonstrate the error of Pascal and Borelli, who thought humid air to be heavier than dry air. He considered experiments with capillary tubes done by the Accademia del Cimento and by Robert Boyle. Variations on them proved that the rise of water was due to viscosity and not to the supposed difference, alleged by Honoré Fabri, between the quantity of air pressing upon the cistern and the quantity of air that could squeeze into the tube's upper opening. Otto von Guericke's invention of the air pump, with the improvements made upon it by Boyle and Robert Hooke, permitted still further experiments.[75] A sectional diagram showed how the simplest models functioned. Other experiments demonstrated the elasticity of air and the falsehood of Aristotle's *horror vacui*—at least in the absolute sense no longer tenable in contemporary physics. The air thermoscope of Galileo and Santorio Santorio demonstrated why *experientia repugnat* any other hypothesis for explaining the fall or rise of the liquid than the communication of heat or cold in the surrounding atmosphere to the air in the bulb.[76] Even students with no interest whatsoever in becoming scientists must have found both enjoyment and instruction in seeing this popularized experimental philosophy at work.

On the basis of simple conclusions drawn from these demonstrations, Poleni led his students on a tour of modern theories about matter and motion. He defended Descartes' purely mechanistic physics from Gassendi's attempt to solve many of the problems associated with the concept of motion by infusing the mechanical world with a general hylozoic sensitivity. To Gassendi's use of a biological metaphor for explaining the circulation of vapors in the earth, he responded, "Those philosophers who ineptly and impiously supposed the Earth to be a kind of giant animal, cannot, to my mind, have found any real analogy between the bodies of animals and that of the earth, except to the extent that the interior parts of both animals and the earth are perpetually heated."[77] He drew his explanation of the structure of matter straight from the *Principia Philosophandi* and *Les météores*. To show how the sun communicated a whirling motion to the elongated particles of water issuing from the earth as rising vapor, he went to the blackboard with the diagram from *Les météores* of two strings flying outward from a spinning pole. Just in case this was not enough to explain the effect, he adopted the additions of Pierre Sylvan Régis, who suggested that the emerging particles bore an outer layer of very subtle matter, so the increasing diameter of the sphere they described by their heat-induced spinning would reduce their density with respect to this outer layer, allowing it to carry them up to the rarer parts of the atmosphere.[78] Moving on to the objections of Leibniz to Newton, he asserted that the void, and Newton's concept of space as *sensorium divinitatis,* seemed an unnecessary—and probably dangerous—metaphysical notion.[79]

Poleni had no qualms about straying from the strict Cartesians when it suited his purpose. In hydrostatics, for example, he followed Louis Carré, Malebranche's right-hand man in the Académie des Sciences, in asserting a cautious proto-Newtonian theory of the attraction of certain particles to one another. No one could deny, he noted, that water tended to adhere to substances, and this favored its movement in one direction rather than in another under certain circumstances. He thus pointed the way clearly for the doctrine of attraction asserted by Francis Hauksbee, the Newtonian experimenter, in connection with capillary action.[80] To explain cohesion in matter he followed another Malebranche ally, Jacob Bernoulli (now better known for the *Ars conjectandi* than for the mechanical works). Like Bernoulli and many other contemporaries, he was not entirely satisfied with Descartes' principle of relative quiet—the state implied in the law of inertia—as an explanation for why the particles of an object stuck together. He therefore welcomed the suggestion that the pressure of the aether, accompanied by that of the air itself, was sufficient to account for the rigidity of even the hardest metal.[81]

However, bringing in his own instruments, at great inconvenience and expense, or providing descriptions of experiments that might be so much more illustrative if performed in class, eventually became a burden. Finally in 1719 Poleni proposed a new laboratory for experimental philosophy. Like the recently opened laboratory of experimental philosophy at the Institute of Bologna, it would provide a point of contact between professors' professional interests and the interests of the best students. At first, all that would be necessary would be a few key machines for experiments in hydrostatics, phonics, aerometry, optics, chemistry, and the science of, as he put it, "the bodies that have the quality of attracting each other."[82] In a frank appeal to the well-known parsimony of the Riformatori to whom he addressed the proposal, he included a list of expenses for the first year down to the last ducat, and a full account of the benefits to the university. "New discoveries," he pointed out, "enlarge the minds of those who make them and are a great object of excitement for those who observe them."[83] Foreign students whom wars, pestilence, and the convenience of local institutions had long discouraged from attending, would surely return in great numbers. However, instead of welcoming Poleni's project, the Riformatori brushed it aside just as their predecessors had done with the similar projects presented by Domenico Guglielmini and Pompeo Sacco at the end of the previous century and by Scipione Maffei in the report on university reform in 1715.[84] In 1724, they accepted Poleni's second and much more modest proposal: the acquisition of a set of brass interlocking squares, circles, and triangles. That way, at least, he might be able to execute his mathematical diagrams without having to choke on the clouds of dust he usually raised by drawing things freehand on the blackboard. So pleased were the Riformatori with this new acquisition that they insisted on having his name engraved upon them for display after his death, where they are to this day, in a hall of the Palazzo del Bò.[85]

It was only a matter of time. Soon the popularity of science became impossible to ignore. encyclopedic periodicals with titles like *Giornale de' letterati, Raccolta d'opuscoli,* and so forth, continued to flood the market. A new Venetian edition appeared of a work by Willem Jacob 'sGravesande, whom Poleni claimed had, along with Petrus van Musschenbroek, "turned experiments into an art."[86] In 1738, the Riformatori finally gave in. Granting official recognition to experimental philosophy at Padua, they established a professorship and ordered the outfitting of a laboratory. Over fifty years had passed since similar arrangements had been made at the University of Leiden, and somewhat fewer at Utrecht, Cambridge, the Institute of Bologna, and the University of Turin. Finding a good foreign candidate did not promise to be easy, now that experimental philosophy had come into vogue all over Europe. As Poleni told the Riformatori, "the men who cut the most impressive figures in the Experimental Art these days, and they are very famous, would not without the greatest difficulty (besides the fact that they follow a different religion) be induced to leave their comfortable and lucrative positions."[87] Not by chance, the choice fell on Poleni himself, whose blueprint was to constitute the new course's practical and theoretical foundation.

Poleni's teaching in the new post followed the van Musschenbroek and 'sGravesande approach of translating simple mathematical notions into visual terms. A lesson on the physics of simple machines, for instance, sought not to arrive at new conclusions, but to prove a didactic point. He attached various weights to a block and tackle and hoisted them by a device that permitted measurement of the work done. This experiment, drawn out of 'sGravesande's *Physices,* was supposed to show students that in cases of two or more pulleys, "the ratio of force to weight, when the movement is uniform, is composed of the proportion between the diameters of the last wheel pulling the weight and the first wheel whose circumference is subject to it on the one hand and, on the other, of the proportion between the revolution of the last wheel to that of the first wheel in the same time."[88] Instead of centering upon the broader inferences that could be drawn from the interplay of simple mechanical laws, Poleni focused discussions upon the chaotic world of observable mechanical phenomena. Rather than demonstrating, through the use of mathematical reasoning, that the universe was an integrated and harmonious whole, the most Poleni could hope from this kind of teaching was that his students should shun unfounded hypotheses while learning to regard higher mathematics as something more than a recondite game for intellectual initiates.[89]

Accordingly, over the course of twenty-five years, Poleni had the Riformatori acquire more than 250 machines of all kinds, many of them from German, Dutch, and French shops. There was Musschenbroek's famous pyrometer (first introduced in the *Tentamina* of 1731) consisting of a metal rod fastened down at one end and, at the other, fitted with gears and wheels to measure expansion when heated.[90] There was a heliostat, another specialty from the Musschenbroek shop, which consisted of a mirror fixed to a clock permitting the operator to study the sun as it passed across the sky, without having to readjust the apparatus continuously.[91] There was a machine whose "construction and the motion of its parts are such that various bodies can be made to rotate in various ways for exploring the properties of centrifugal force."[92] Perhaps it was one of the Musschenbroek models, with a bell that rang when the weight on the spinning bar had slid out beyond a certain point. There was a crank-operated electrical machine with "various cylinders and various globes of glass or crystal," probably activated by the friction of the operator's hand. It featured "two plates with little chains and tubes, all of iron," apparently a variation of Georg Matthias Bose's recently proposed apparatus for demonstrating the conductivity of metal.[93] There was a set-up involving a "large wooden pole for holding the scale-pan of a balance. Conjoined is another iron for holding the beam of the balance. Cylindrical scale-pans and balls of various weights."[94] In the inertia and impact experiments mentioned by 'sGravesande using this device, weights were tied by a string and allowed to fall into the pan at one end of the balance in order to show how much liquid had to be poured into containers in the pan at the other end for equilibrium to be maintained.[95] Then there were the usual air pumps and their attachments for such demonstrations as that old textbook favorite, the equal fall of a lead ball and a feather in an evacuated jar.[96] Telescopes of various lengths included a two-foot *telescopium catoptricum dioptricum sive Newtonianum*.[97] Much less clear is the possible use of a "recent invention" for investigating "the pores of bodies" and consisting of "four closed-off metal capsules with a wooden instrument inserted along which they can slide."[98]

In perfecting the new teaching method, Poleni encountered all the usual problems of a first-time lecturer. "I was disturbed by the insurmountable inconvenience of the locale," he wrote to his fellow-scientist in Venice, Giovanni Francesco Pivati, "by the crowd, by the heat, and by the difficult task of saying and performing while trying to make sure that both should come out pleasing and correct."[99] He hoped that with continuous practice of "saying the things in the lessons that will subsequently be done during the demonstrations for listeners who will also be the spectators in the experiments" he would eventually find "the system most adapted to the purpose."[100]

Elegant presentation was particularly important considering the extra duties Poleni was expected to assume, of entertaining the various visitors who came through Padua to admire the handiwork of the Riformatori. To "Il Signor Capitano Venier and Dame Correr," in 1753, for instance, Poleni showed "the double cone that ascends an inclined plane," "the center of gravity, with the little statue," "the friction machine of Signor [Jean Antoine] Nollet," and, his own invention, "the machine for extinguishing fires."[101] A master list of "experiments done" reminded him of those that were "good" and those that were not. One of his best, which not only demonstrated a fundamental biological principle but also performed a useful service, no doubt winning him the applause of the Venetian noblemen and noblewomen for whom he performed it, was listed as "with the air pump, death of a pigeon."

By a hair's breadth, Poleni managed to avoid turning experimental philosophy into a sort of sophisticated mountebank show, such as those afforded by Bose in Wittenberg, with the "wonderful displays" of electrical fluid, full of surprises for his awe-struck public, or John Banks in Manchester, with his half-crown lessons on such useful subjects as "the application of water to Bucket Wheels," which imparted, according to one listener, "a wonderful insight into the laws of nature."[102] Not to mention the various performers who showed up at fairs and public houses. Instead, Poleni hoped to emulate the widespread and successful version of experimental philosophy preached by Jean Théophile Desaguliers at Mr. Brown's, Bookseller, in London or by Martinus Martens in various locales around Amsterdam.[103] To keep the more serious purpose of his efforts in mind, he prepared laboratory notes for his own and others' future use. "I think that in this way a useful, clear, and scientific order would arise by itself, from which a completely and entirely regular course could be constructed according to an exact method."[104] His success in establishing such a system was attested by honorable mention along with Desaguliers, Daniel Bernoulli, Christian Wolff, and Musschenbroek among the great experimenters of the age in the preface to the third edition of 'sGravesande's *Physices* (1742) even though he never published a major compendium of his experimental work.

Thus the noblest purposes of the university reformers were fulfilled in the teaching of science at Padua. Not only did Vallisneri and Poleni manage to adapt their teaching methodologies to the needs of their public by convincing the university administration to make the necessary modifications, but they brought to their lessons some of the scientific expertise that had won them prizes and accolades from colleagues and academies abroad. They knew that most students had no interest in broadening and deepening the perspectives gained in the course. They found consolation in the few individuals in the crowd who showed signs of sudden illumination and came back to take private lessons. Not to mention the satisfaction derived from public acclaim. And a sufficient number of students considered the profession to be attractive enough to abandon more lucrative ones, contributing to the complicated momentum that Paduan science had continued to maintain, in spite of whatever metaphysical problems it encountered, even in periods when it was eclipsed by the appearance of greater genius—and, no doubt, more abundant publishing opportunities—elsewhere.[105]

In spite of Cesi's fears about the deleterious effects of the marketplace, the future clearly belonged not only to the most skillful teachers but to researchers whose published work somehow struck a chord in the public imagination. Not surprisingly, when the *Giornale de' letterati d'Italia*, run by Apostolo Zeno and Scipione Maffei in Venice from 1710, the most important general knowledge periodical in Italy, offered to Vallisneri and Poleni the possibility of publishing anonymous favorable reviews of their own work in the journal, just as the *Galleria di Minerva* had done, they leapt at the opportunity.[106] What better way to silence adversaries and ensure that their own ideas would receive a better reception among patrons and public alike! By the second half of the eighteenth century, even the erstwhile enemies of innovation in natural philosophy appealed to the public to support their cause, in a remarkable reversal of roles that we shall analyze in the following chapter.

Notes

1. ASV, *Riformatori,* filza 409, Antonio Bombardini, 8 December 1717: "Al pregiudizio dell'economia paterna, s'aggiunge la distrazione dello scolare. Viene questo allo Studio con la certezza di dovere tornare per quattr'anni. Questo riflesso fà che il primo anno appena appena imparano a conoscere i professori di vista per salutarli per strada, applicando a tutto ciò che meno conviene allo studio, e questo habito fatto nel primo anno continua fatalmente fino all'ultimo anno del dottorato, in cui abbattutisi come le vergini pazze del Vangelo senza olio nel lume, corrono freneticamente dai Professori con l'inopportuno: date nobis de olio vestro."

2. ASV, *Riformatori*, filza 197, Bortolo Sellari, 7 November 1723: "È più costante che mai l'opinione che lo Studio si veda in un'estrema miseria in quest'anno. Si legge, ma alla riserva di pochi greci, gli altri non sono che gli amici dei Pubblici Professori."

3. Federico Cesi, "Del naturale desiderio di sapere," in Ezio Raimondi, ed., *Narratori e trattatisti del Seicento* (Milan-Naples: Ricciardi, 1960).

4. BNC, Magl. cl. VIII, cod. 1089, fol.7r (10), Antonio Vallisneri to Antonio Magliabechi, 17 July 1700: "Ho fatto assai ad attendere al grave peso delle pubbliche lezioni, senza divertirmi a ricercare le operazioni più curiose e più astruse della natura. Non potevo scrivere, che cantilene da farsi sopra le cattedre, e gloriosi inganni dell'Arte nostra. Ora torno a dare un'occhiata a vecchi studi, dilatandoli a quanto di grande e di rimarcabile sa operare la Natura."

5. On the sixteenth-century reform of universities, see Paul Oskar Kristeller, "The University of Bologna and the Renaissance," in *Studi e memorie per la storia dell'Università di Bologna*, 2 vols. (Bologna: Istituto per la storia dell'università, 1956), 1: 313-23; Charles B. Schmitt, "Science in the Italian Universities in the Sixteenth and Early Seventeenth Centuries," in *The Emergence of Science in Western Europe*, ed. Maurice Crosland (London: Macmillan, 1976), 47-50; and Eric Cochrane, *Italy 1530-1630*, ed. Julius Kirshner (New York: Longmans, 1988), ch. 3: "The New Political Order."

6. Charles B. Schmitt, "The Faculty of Arts at Pisa at the time of Galileo," *Physis* 14 (1972): 263. With particular regard to the Paduan period, compare Antonio Favaro, *GalileoGalilei e lo Studio di Padova*, 2 vols. (Florence: Le Monnier, 1883), 1: 168; Ludovico Geymonat, *Galileo Galilei* (Turin: Einaudi, 1969), 34. For my information on Malpighi I used Howard B. Adelmann, *Marcello Malpighi and the Evolution of Embryology*, 5 vols. (Ithaca: Cornell University Press, 1966), 1: 157.

7. *Circulus pisanus Claudii Berigardii Molinensis, olim in Pisano, iam in lyceo Patavino Philosophi primi paris. De veteri et peripatetica philosophia* (Udine: Nicolo Schiratti, 1643), 16. I cite from the first edition, emitted in 3 separate parts: *In priores libros physici Aristotili; In octavum librum physicorum Aristotili; De ortu et interritu*, instead of from the revised and elaborated later edition of 1661-62, since the earlier one more clearly reflects his Pisan debates in the *circoli* in spite of the invented interlocutors. He is principally known for his opposition to Galileo's cosmography, which is the basis for recent studies on him by Maria Laura Soppelsa, *Genesi del metodo galileiano e tramonto dell'Aristotelismo nella scuola di Padova* (Padua: Antenore, 1974), 92-112; and Giorgio Stabile, "Il primo oppositore del *Dialogo*: Claude Bérigard," *Novità celesti e crisi del sapere*, Atti del Convegno Internazionale di Studi Galileiani, marzo 1983, ed. Paolo Galluzzi (Florence: Giunti-Barbèra, 1984) 277-82.

8. *Circulus pisanus: In priores libros physici Aristoteli*, 17.

9. *Circulus pisanus: In librum 2 de anima Aristoteli*, 63-67; *Circulus pisanus: De ortu et interritu*, 104-07.

10. Portions of Montanari's lessons, which exist in BCV, in a manuscript entitled *Trattati di matematica e di fisica composti e dettati dal Sig. Dottore G. M. Pubblico Porfessore di Meteore nella Università di Padova e scritti da me Francesco Bianchini in quella città negli anni 1682, 1683,* have been published by Salvatore Rotta, in appendix to "Scienza e "pubblica felicità" in Geminiano Montanari," *Miscellanea Seicento*, Istituto di filosofia della facoltà di lettere e filosofia dell'università di Genova, vol. 2 (Florence: Le Monnier, 1971), 187-99. Says Rotta, 205: these manuscripts "possono . . . tener luogo degli autografi perduti."

11. *Trattati di matematica e di fisica*, 191.

12. *Trattati di matematica e di fisica*, 192.

13. *Trattati di matematica e di fisica*, 193.

14. Michael Heyd, *Between Orthodoxy and the Enlightenment: Jean-Robert Chouet and the Introduction of Cartesians Ceince in the Academy of Geneva* (The Hague: Nijhoff, 1982); Wolfgang Rother, "Zur Geschichte der basler Universitätsphilosophie im 17 Jahrhundert," *History of Universities* 2 (1982): 169; Laurence W. B. Brockliss, "Aristotle, Descartes and the New Science: Natural Philosophy at the University of Paris, 1600-1740," *Annals of Science* 38 (1981): 46; Edward G. Ruestow, *Physics at Seventeenth and Eighteenth-Century Leiden. Philosophy and the New Science in the University* (The Hague: Nijhoff, 1973), 61-72; L. C. Palm, "Snellius and his Newtonian Teaching in Halle," *Janus* 64 (1977): 15-24.

15. Thomas Willis, for example, incorporated his Oxford lectures on the functions of the brain and neurological diseases in *De anima brutorum* (1672). *Thomas Willis' Oxford Lectures*, ed. and trans. Kenneth Dewhurst (Oxford: Sandford Publications, 1980).

16. Salvatore Rotta, "Scienza e 'pubblica felicità'," 98.

17. Concerning Sbaraglia, see Adelmann, *Marcello Malpighi*, 1, chap. 7; Ugo Baldini, "La teoria della spiegazione scientifica a Bologna e Padova," in *Rapporti tra le università di Padova e Bologna. Ricerche di filosofia medicina e scienza*, ed. Lucia Rossetti (Trieste: Edizioni Lint, 1988), 223-24, 240-41.

18. Vallisneri's notes exist in MCP, ms. C.M. 536, entitled *Sylva medico-physica—De pulsibus*, fol. 76r: "Sed hic placet adnotare velle methodum cognoscendi veritatem vel falsitatem alicuius asserti, . . . an facilius solvantur difficultates in una opinione, quam in altera."

19. *Sylva medico-physica*, fols. 74v, 77r.

20. *Sylva medico-physica*, fol. 23r: "Denudent arteriam, et digitis comprimant eamdem singulis pulsibus; quando sanguis in arteriae partem descendentem pertransibit, pulsus apparebit idem et diversus, prout maior, vel minorem copiam sanguinis progredi permittent."

21. *Sylva medico-physica*, fol. 22v: "Fracastorus voluit pulsum a facultate insita ipsis arteriis fieri. Galenus autumavit pulsum fieri a corde influente in tunicas arteriarum facultatem pulsificam, probavitque hanc suam sententiam in Lib. An [*in arteriis natura*] *sanguis* [*contineatur*], experimento calami in arteriam immissi, dein ligata arteria, sub qua ligatura arteriam moveri prodidit, licet sanguis per cavum calami transiret. . . . Neque quicquam probat experimentum . . . etenim infra ligaturam immisso calamo adhuc observatus pulsus, ut EGO SUM EXPERTUS [capitals in original] quicquid alii commentarentur, circa propositum experimentum, et quando pulsus non micat infra ligaturam est quia sanguis non transit grumefactus per cavum calami." The demonstration in question is from Galen, *An in arteriis natura sanguis contineatur* in Galen, *Opera*, vol. 1 (Venice: 1586), fol. 62v.

22. Information on Cucchi is in Serafino Mazzetti, *Repertorio di tutti i professori antichi e moderni della famosa Università . . . di Bologna* (Bologna: Tip. di S. Tommaso d'Aquino, 1847), 103, 224; Jean-Pierre Niceron, *Mémoires pour servir à l'histoire des hommes illustres*, vol. 14 (Paris: Briasson, 1731), 224. Vallisneri's impression of Malpighi is in Giovan Artico di Porcia *Notizie della vita e degli studi del Kavalier Antonio Vallisneri*, ed. Dario Generali (Bologna: Patrón, 1986), 47. In fact, this work is essentially Vallisneri's autobiography.

23. *De recentiorum medicorum studio* (Göttingen: 1687 [but Bologna: 1689]), reprinted in Sbaraglia, *Exercitationes physico-anatomicae* (Bologna: P. Monti, 1701), 20: "Aliud . . . est anatomiam studere ad penitissimam structuram cognoscendam, aliud ad therapiam in ultionem morborum administrandam" and "Omnino aliud est philosophari, aliud mederi."

24. I consulted the *Institutiones Medicae* in Lazare Rivière, *Opera Omnia* (Venice: Baglioni, 1735), chap. 1.

25. *Sylva medico-physica*, fol. 7v.

26. *Sylva medico-physica*, fol. 10r: "Non [*potest*] negari usus cordis in ordinem ad elaborationem sanguinis, quod idem suadet structura cordis, quae non est instituta a natura tantum ad motionem; et hoc deducitur ex causa finali circulationis, quam recentiores adducunt, scilicet ut sanguis crudus ad cor regrediatur, ut denuo elaboretur, et profiteatur."

27. *Sylva medico-physica*, fol. 8r. Sbaraglia mentions in particular also the Dutch physiologist Frans de Le Boë (Franciscus Sylvius).

28. *Sylva medico-physica*, fol. 10r: "Sed dicant quicquid velint, placet a structura heppatis, [*atque*] a circulatione sanguinis, ad nullum finem apparentem spiritum naturalem excludere. [*Excludendus est spiritus naturalis*] a structura heppatis, quae docet heppar formatum esse a natura, ut a sanguine continuo secernatur biles excrementitia, [*atque*] a sanguinis circulatione, nam haec demonstrat per venas non posse quicquam corpori dispensari, a nullo fine apparente: partes enim a sanguine et succo nerveo sufficienter irrigantur pro quolibet suo munere." According to Rivière, *Institutiones Medicae*, 12, "[*Spiritus*] Naturalis in iecore producitur ex tenuiore sanguinis portione, cum pauco aere permixto, et per venas influit in totum corpus, ut naturales facultates debite exerceantur." Santorio Santorio, *Commentaria in Artem medicinalem Galeni* (Venice: 1612), 257, had denied the existence of the natural spirit in, "An detur spiritus naturalis praeter animalem et vitalem." On this topic, Nancy G. Siraisi, *Avicenna in Renaissance Italy* (Princeton: Princeton University Press, 1987), 340-42.

29. *Sylva medico-physica*, fol. 10r: "Vitalis concedendus est, sed non genitus ex aere et spiritu naturali, verum est defaecatior, purior, et subtilior sanguinis portio activissima in corde soluta et exaltata, cuius signum est quando sanguis aggreditur a corde, tenuior est, etc." The phrase of Michael Servetus, *Christianismi Restitutio* (Vienne [Dauphiné]: B. Arnollat, 1553), lib. 5, 168, is cited by Walter Pagel, *New Light on William Harvey* (New York: Karger, 1976). Galen's explanation is in *De Usu Partium*, parag. 539-42, consulted in the edition in *Opere scelte*, ed. Ivan Garofano and Mario Vegetti (Turin: UTET, 1977), 499-500.

30. *Sylva medico-physica*, fol. 10r. Harvey's words are in *Exercitationes de generatione animalium* (Padua: Frambotto, 1666), Exercitatio 70, "De calido innato," 500.

31. Porcia, *Notizie*, 52. Concerning Vallisneri, there is Paola Masat Lucchetta, *A. V. medico naturalista: Scienza e filosofia nel Settecento* (Venice: Cafoscarina, 1984); and Mario Sabia, *Le opere di Antonio Vallisneri: medico e naturalista reggiano, 1661-1730: bibliografia ragionata delle opere vallisneriane* (Rimini: Luise, 1996). Still somewhat useful are the articles in *Il Metodo sperimentale in biologia da Vallisneri in biologia da Vallisneri ad oggi; III centenario della nascita di Antonio Vallisneri, 29-30 settembre-1 ottobre 1961* (Padua: Accademia patavina di scienze lettere ed arti, 1962).

32. At a certain point, where he had written "Terapia ex Sbaralea" he cancels the name of Sbaraglia and puts his own. *Sylva medico-physica*, fol. 43r.

33. *Sylva medico-physica*, fol. 29r, text of Sbaraglia, with additions by Vallisneri underlined: "Convulsivus ortum habet ab inflammatione, et convulsione fascicolorum nervosorum et lacertorum cordis."

34. *Sylva medico-physica*, fol. 60r: "Concludamus igitur quod facultas pulsifica satis habuit gloriae, satis per tot saeculi credulo Tyronum vulgo imposita [*fuit*]. Excussum est, N[obiles] I[*uvenes*], tyrannicum illud iugum, quod debeamus solum ex commentario sapere."

35. For example, Ibid., fols. 33r: "Alumen pulverisatum nullam perturbationem parit urinae recenti et tepidae, sed frigidae aliquam excitat perturbationem, ita ut sedimentum crassum fundum petat. . . . Aqua fortis urina immixta magnam effervescentiam excitat, et praecipitat materiam albam."

36. Ibid., fol. 22v: "Laudentur antiqui patres . . . sed magis laudetur experientia, laudetur veritas a cimeriis eruta tenebris et caliginoso denudata flammeo."

37. Porcia, *Notizie*, 77: "La maniera d'insegnare del nostro Vallisneri si era questa. Spiegava prima il testo, sopra quale dovea quel tale anno ragionare, fosse d'Ippocrate, di Galeno, o di Avicenna." A curious feature of the documents of the first lessons is that the ancient text is mentioned and indicated various times, but never cited at any length. From this I deduce that the text was recited in the beginning of the cycle and omitted from the rest of the lessons.

38. Porcia, *Notizie*, 73: "Febris est calor extraneus accensus in corde et procedens ab eo mediantibus spiritu et sanguine per arterias et venas in totum corpus." The text is drawn from the translation by Cesare Cremonini of the *Liber Canonis* (Venice: Paganini, 1507), lib. 4, fen 1, cap. 1. This account is the only testimony of the first year of Vallisneri's lessons, since the manuscripts have not survived. Compare Roberto Savelli, "L'opera biologica di A. V.," *Physis* 3 (1961): 291; and Adalberto Pazzini, "Vallisneri e Morgagni nel magistero di Padova," in *Il metodo sperimentale in biologia da Vallisneri ad oggi* (Padua: Accademia patavina di scienze, lettere ed arti, 1962), 119.

39. The lessons on head diseases have survived in a manuscript in PBA, ms. 1796, entitled "Lectiones." The following observations are based on this manuscript, henceforth denoted as *Lessons (1701-2)*.

40. "Soda [*i.e., dolor capitis*] est lesio in membris capitis. Omnis autem laesionis causa est mutatio complexionis: aut eius diversitas; aut est solutio continuitatis, aut aggregatio amborum simul," from *Liber Canonis*, lib. III, fen 1, tract. 1, cap. 2.

41. *Lessons (1701-2)*, fol. 26v: "Hanc postea non feci ne nimia vermium frequentia taedio afficeret."

42. *Lessons (1701-2)*, fol. 3r: "Non sumus enim adeo recentibus addicti, et emancipati, ut quaelibet eorum commenta clausis oculis sequi velimus."

43. Porcia, *Notizie*, 74: "Diede quindi un'altra definizione rispondente alle nuove osservazioni, alla circolazione, ai movimenti, e alle quantità del sangue, e incominciò a promulgare dottrine opposte alle antiche . . . dimostrando i fenomeni de' corpi animati dipendere dall'equilibrio fra i fluidi, e i solidi, dai movimenti regolati, o sregolati de' medesimi, dalla elasticità, dalla pressione." The description, evidently interpolated by Porcia, appears inexact.

44. Venice, Biblioteca Nazionale Marciana (=BNM), lat. X: 148 (6685) [henceforth, *Lessons (1710-11)*], fols. n.n., lezione 23: "Testes hominum esse partem principem ratione speciei, quam confirmatur ex textu Galeni quem explicandum sumpsimus et ex nostris observationibus, in antecedentibus scholis didiscistis." I compared *Ars medica*, cap. 9: "Quot sint differentiae partium" (in *Opera*, I, 63) with *De semine*, lib. 1 cap. 15 (in *Opera*, I, 334).

45. Vallisneri changed his view with regard to Leeuwenhoek's preformationism about five years after beginning these lessons, as we see from a letter to Ludovico Antonio Muratori in February, 1715, edited by M. Nichetti Spanio in L. A. Muratori, *Carteggi con Ubaldini . . . Vannoni*, Edizione Nazionale del Carteggio di L. A. Muratori, vol. 44 (Florence: Olschki, 1978), 191.

46. *Lessons (1710-11)*, lezione 37: "Pythagorae discipuli tanto pendebant [atque] adeo venerabantur sui praeceptoris oracula ut si quando affirmarent in disputando, cum ex his quaereretur, quare ita esset, respondere solebant 'Ipse dixit,' ipse autem erat Pythagoras. Nolo vos simili authoritati magistri fore ita alligatos, nec enim ego sum Pythagoras, nec vos istae viles illorum Graeculorum discipulorum animae. Volo ut ponderetis meas rationes solum ut dicatis non 'Ipse dixit,' sed 'Ipsa dixit natura'."

47. Cited by Giuseppe Gullino, "Un eredità di consigli e di salutari avvertimenti: l'istruzione morale, politica ed economica di un patrizio veneziano al figlio (1734-38)," in *I ceti dirigenti in Italia in età moderna e contemporanea*, Atti del Convegno, Cividale del Friuli, 10-12 settembre 1983, ed. Amelio Tagliaferri, Serie monografica di Storia moderna e contemporanea dell'Istituto di Storia dell'Università di Udine, 8 (Udine: Del

Bianco, 1984), 339.

48. *Lessons (1701-2)*, fols. 22v, 79r-80r, 87v, 88v.

49. Biographical details are drawn from the anonymous life in BNM, cod. ital. cl. XI: 220 (= 6940) datable to just after Poleni's death in 1761. Also, Cossali, *Elogio*; Giuseppe Gennari, *Elogio di G. P.* (Padua, 1839). On his scientific contribution, summarized in the entry by Bruno A. Boley in DSB, 11 (1975), 65-66, see the articles in the collection, *G. Poleni (1683-1761) nel bicentenario della morte* (Atti e memorie dell'Accademia Patavina di Scienze, Lettere ed Arti, 74, Padua, 1961-62, special number); and, in addition, P. Verrua, "La macchina calcolatrice di G. Poleni," *Bollettino dell'Accademia Italiana di stenografia* 7 (1931): lxix-lxxvi; Frances Vivian, "Joseph Smith, G. Poleni and Antonio Visentini in the Light of New Information Derived from the Poleni Papers in the Marciana Library," *Italian Studies* 18 (1963): 54-66; L. Guadagnino Lenci, "Per G. Poleni: note ed appunti per una revisione critica," *Atti dell'Istituto Veneto di scienze, lettere, ed arti* 134 (1975-76): 543-67; Augusto Cavallari-Murat, "Collaborazione Poleni-Vanvitelli per la cupola vaticana, 1743-8," in *Luigi Vanvitelli e il Settecento Europeo: congresso internazionale di studio, Napoli-Caserta, 5-10 dicembre 1973* (Naples: Università, Istituto di storia dell'architettura, 1979), 1: 171-210; Idem, "Le *Exercitationes Vitruvianae*: approdo neoclassico di Simone Stratico (1733-1824)," *Filosofia* 29 (1978): 445-76; Silvio Bergia and P. Fantazzini, "Della regolamentazione delle acque alle leggi dell'urto centrale elastico: una rivalutazione del ruolo di G. Poleni," *Giornale di fisica* 21 (1980): 46-60; Dante Nardo, "Scienza e filologia nel primo Settecento padovano: gli studi classici di Giovanni Battista Morgagni, Giovanni Poleni, Giulio Pontedera e Leone Targa," *Quaderni per la storia dell'Università di Padova* 14 (1981): 1-40.

50. Ugo Baldini, "L'attività scientifica nel primo Settecento," in *Storia d'Italia*, Annali 3: *Scienza e tecnica* (Turin: Einaudi, 1980), 509 with bibliography.

51. In Venice, Biblioteca Nazionale Marciana. Precise indications will be given in due course.

52. On the Collegio alla Salute, see Giannantonio Moschini, *La Chiesa e il Seminario di S. M. della Salute in Venezia* (Venice: G. Antonelli, 1842), 53-147; Silvio Tramontin, "Gli inizi dei due Seminari di Venezia," *Studi veneziani* 7 (1965): 313-77; Giuseppe Gullino, *La politica scolastica veneziana nell'Età delle Riforme* (Venice: Deputazione di storia patria, 1973), 76; 107.

53. There is no study of Caro. On the Somaschi program, Marco Tentorio, "I Somaschi," in *Ordini e congregazioni religiose*, ed. Mario Escobar (Turin: Società editrice internazionale, 1951), 626; Francesco De Vivo, "Indirizi pedagogic) ed istituzioni religiosi nei secoli 16 e 17," *Rassegna di pedagogia* 16 (1958): 1-23.

54. Francesco Caro, *Philosophia amphysica*, 5 vols. (Venice: Alvise Pavino, 1695), 3: 6.

55. *Ibid.*, 47. Nicolas Cabeo is remembered rather for his experiments on electrical repulsion.

56. *Ibid.*, 1: 155.

57. *Ibid.*, 3: 6. Consider Clelia Pighetti, *L'influsso scientifico di Robert Boyle nel tardo Seicento italiano* (Milan: Angeli, 1988).

58. On Maffei, see *Giornale de' letterati d'Italia* 28 (1717): 387; *Scriptores Ordini Predicatorum* 3 (1910), 258.

59. On the Venetian government's cultural policy, see B. Dooley, "Giornalismo, accademie e organizzazione della scienza: tentativi di formare un'accademia scientifica veneta all'inizio del Settecento," *Archivio veneto*, ser. 5, vol. 120 (1983): 5-39.

60. See Conti's letter to Poleni, BNM, cod. lat. cl. VIII:136 (=2147), 9 July 1728.

61. BNM, cod. lat. cl. VIII: 147 (=2724), n.p.: "Ex observationibus rationi maxime consentaneum extitit supponere quodcumque sidus peculiari atmosphaera non secus ac tellurem nostram circumdari."

62. BNM, cod. lat. cl. VIII: (=2724), n.p.: "Ipse celeberrimus Newtonus, qui admirabili sublimis ingenii pompa caelestium spatiorum materiam tam sedulo exinanuit, censit tamen, quod cometae cauda nihil aliud sit, quam vapor longe tenuissimus, quem caput, seu Nucleus cometae per calorem amittit."

63. On Andrea Argoli, see entry by Mario Gliozzi in DBI 4 (1962): 132-34; and Soppelsa, *Il metodo galileiano* 70-79. On Degli Angeli, see Luigi. Tenca, "Stefano Degli Angeli," *Attidell' Accademia delle. Scienze. dell'Istituto. di Bologna*, ser. 2, 5 (1958): 194-207.

64. The lessons on astronomy for the academic year 1714-15 are in BNM, cod. lat. cl. VIII: 157 (=2734), (hereafter, *Lessons 1714-15*, I, numbered consecutively). Lessons will be indicated in the notes by these numbers.

65. Salvatore Rotta, "Sulla costruzione e diffusione in Italia dei telescopi a riflessione," *Le macchine* 1 (1968): 97-98.

66. See Bernoulli's letter in BNM, cod. ital. cl. IV: 642 (=5503), 26 April 1726.

67. ASV, *Riformatori* filza 83, 30 January 1720-21.

68. *Lessons, 1714-15*, 1, n. 37. Compare Johannes Hevelius, *Cometographia* (Gdansk: Simon Reiniger, 1668), an obvious source.

69. *Lessons, 1714-15*, 1, n. 38: "Coelestium rerum periti duplici usi sunt methodo pro tali determinatione assequendo qua geometricis et arithmeticis, qua opticis theorematibus insistendo."

70. *Lessons, 1714-15*, 1, n. 44. See André Tacquet, *Opera, 1. Astronomia* (Antwerp: Meursium, 1669), Bk. 8, chap. 2. On Tacquet's "annoyingly erudite eclecticism," see the entry by P. Stromholm, DSB 13 (1976), 235-36.

71. *Lessons, 1714-15*, 1, n. 47: "[Deus] infinita sapientia sua hanc naturae machinam ita ordinavisse ut causae secundae previa causae primae ordinatione agant."

72. Some of his figures are in *Lessons, 1714-15*, 2, fols. 119-22. They are in multiple copies on squares of paper identical in size, probably made up for distribution to students.

73. *Lessons, 1714-15*, 2, fol. 75. To place these pneumatic experiments into the context of the history of technology I used William E. Knowles Middleton, *History of the Thermometer* (Baltimore: Johns Hopkins, 1966), chaps. 1-21; Idem, *The Invention of the Meteorological Instruments* (Baltimore: Johns Hopkins, 1969), 1-29, 43-63, 70-71.

74. *Lessons, 1714-15*, 2, fols. 103-04.

75. *Lessons, 1714-15*, 2, fols. 105-06.

76. *Lessons, 1714-15*, 2, fol. 128.

77. *Lessons, 1714-15*, 2, fol. 86: "Philosophi illi, qui inepte, atque impie Tellurem magnum quoddam animal esse finxerunt nullam crediderim inter animalis et Telluris corpora analogiam invenire potuisse, nisi hanc, quod et animalis et telluris interiores parses calve iugi foveantur." Gassendi's simile is in *Opera*, 6 vols. (Lyons: 1658-75), 1:556-57. On this and other aspects of Gassendi's thought I have followed Kurd Lasswitz, *Geschichte der Atomistik vom Mittelalter bis Newton*, 2 vols. (Hamburg-Leipzig: L. Voss, 1890), 2:140-67; Tullio Gregory, *Scetticismo ed empirismo. Studio su Gassendi* (Bari: Laterza, 1961), 179-250; Olivier Bloch, *La philosophie de Gassendi: Nominalisme, materialisme et metaphysique* (The Hague: Nijhoff, 1971), 155-284; Barry Brundell, *Pierre Gasendi: from Aristotelianism to a New Natural Philosophy* (Boston: D. Reidel, 1987).

78. *Lessons, 1714-15*, 2, fol. 97: "Haec vero materiae subtilis adhaesio tamquam aptissima ad explicandas vaporum, et elevationes, et suspensiones predicatur, tum quia revere, hypothesi admissa, res apte exprimitur, tum quia semel admissa vaporum suspensionem propter talem causam, ipsorum etiam descensus quam faciliter explicatur." Poleni cites Régis' *Système de philosophie*, 5 vols. (Lyons, 1691), 3: 235ff. On Régis, Richard A.

Watson, *The Downfall of Cartesianism 1673-1712* (The Hague: M. Nijhoff, 1966), 75-81.

79. *Lessons, 1714-15*, 2, fol. 123: "Vir ille eximius coelestem omnem materiam exinanire contendit in opere illo, quod (liceat ita loqui) redolet quid plus quam humanum, at quod spectat hoc de coelesti vacuo placitum videtur humanum." On the position of Leibniz and his followers, Alexandre Koyré, *From the Closed World to the Infinite Universe* (Baltimore: Johns Hopkins University Press, 1957), Ch. II; A. Koyré and I. Bernard Cohen, "The Case of the Missing *Tamquam:* Leibniz, Newton, and Clarke," *Isis* 52 (1961): 555-66. Poleni's role in the Leibniz-Newton debate in Italy is discussed in Maria Laura Soppelsa, *Leibniz e Newton in Italia. Il dibattito newtoniano, 1687-1750* (Trieste: Lint, 1989), passim.

80. *Lessons, 1714-15* 2, fol. 75. On Carré, see *Oeuvres de Malebranche*, vol. 20: *Documents biographiques et bibliographiques* ed. André Robinet (Paris: Vrin, 1967), 147-53. Compare Hauksbee's conclusions in *Physico-Mechanical Experiments on Various Subjects* (London: J. Senex and W. Taylor, 1719), sec. 5, exp. 8. On early eighteenth-century transformations in Cartesianism, Nicholas Jolley, *The Light of the Soul: Theories of Ideas in Leibniz, Malebranche, and Descartes* (Oxford: Clarendon Press, 1990); Henry Guerlac, "Some Areas for Further Newtonian Studies," *History of Science* 17 (1979): 75- 101; As far as Italy is concerned, the most trustworthy studies are still R. Ajello, "Cartesianismo e cultura oltramontana al tempo dell'*Istoria civile,"* in *Pietro Giannone e il suo tempo. Atti del Convegno di Studi nel Tricentenario della nascita*, 2 vols. (Naples: Jovene, 1980), I, 1-182; and Vincenzo Ferrone, *Scienza, natura, religione: mondo newtoniano e cultura italiana nel primo Settecento* (Naples: Jovene, 1982), 3-108 (translated by Sue Brotherton as *The Intellectual Roots of the Italian Enlightenment: Newtonian Science, Religion, and Politics in the Early Eighteenth Century* [Atlantic Highlands, NJ: Humanities Press, 1995]). Consider also Claudio Manzoni, *Il "cattolicesimo illuminato" in Italia tra cartesianismo, leibnizismo e newtonismo-lockismo: nel primo Settecento (1700-1750): note di ricerca sulla recente storiografia* (Trieste: Lint, 1992).

81. *Lessons, 1714-15*, 2, fols. 123-25. Compare Bernoulli, *Opera Omnia*, 2 vols. (Geneva: Cramer, 1744), 2:106. On the importance of Bernoulli's theory, besides Lasswitz, *Geschichte der Atomistik*, 2:430-35, see Clifford Truesdell, "A Program Toward Rediscovering the Rational Mechanics of the Age of Reason, 1687-1788," *Archive for the History of the Exact Sciences* 2 (1960): 14.

82. ASV, *Riformatori*, filza 195, 23 April 1719: "I corpi, i quali hanno la virtù di attrarsi l'un l'altro non solo sono degni d'osservazione, ma vi è ai nostri giorni qualche sistema famoso di filosofia, il quale dalle attrazioni dei corpi tutto dipende."

83. ASV, *Riformatori*, filza 195, 23 April 1719: "Vorrei porre più in vista ciò che in quel tempo fosse o intorno alla medicina o intorno altre cose più dalla curiosità degli uomini ricercato. E andare poi investigando il modo d'inventare nuove esperienze, perché le nuove scoperte accrescono l'anima a quelli che le fanno, e sono un grande eccitamento a quelli che prima li vedano."

84. The projects of Guglielmini and Sacco were mentioned by Maffei in his *Parere* (1715), published by Biagio Brugi, "Un parere di Scipione Maffei intorno allo Studio di Padova sui principi del Settecento," *Atti dell'Istituto Veneto di Scienze, Lettere, ed Arti* 69 (1909-10): 582.

85. ASV, *Riformatori*, filza 85, Poleni, 2 December 1723.

86. BNM, cod. ital. cl. IV: 592 (=5555), fol. 191: "Questi degli Esperimenti fanno un'arte, e li vanno eseguendo in serie, e con l'ordine ricercato dal sistema della Natura." The Venetian edition was of 'sGravesande's *Introductio* (Venice: Pasquali, 1737). In my remarks on the Dutch tradition in experimental philosophy I follow Giambattista Gori, *La fondazione dell'esperienza in 'sGravesande* (Florence: Nuova Italia, 1972), 7-159.

87. BNM, cod. ital. cl. IV: 592 (=5555), fol. 191, to G. F. Morosini, n.d., but 1738: "Gli uomini che fanno adesso nell'Arte Esperimentale la figura prima, e notissimi sono, troppo difficilmente (oltre che seguono una diversa religione) si indurrebbero a lasciare i loro comodi e lucrosi posti." In general, see G. A. Saladini and M. Pancino, *Il 'teatro' di fisica sperimentale di Giovanni Poleni* (Trieste: Lint, 1987).

88. BNM, cod. lat. cl. VIII: 145 (=2722), "Lezioni, 1743-44," n. 18. Compare Willem Jacob 'sGravesande, *Physices*, 2 vols. (Leyden: Langerak-Verbeek, 1742), 1, Bk. 1, chap. 15.

89. On Newton's methodology, Alexandre Koyré, *Newtonian Studies* (Cambridge, Mass.: Harvard University Press, 1965), chap. 2; I. Bernard Cohen, "Hypotheses in Newton's Philosophy," *Physis* 8 (1966): 163-84; Alan E. Shapiro, *Fits, Passions, and Paroxysms: Physics, Method, and Chemistry and Newton's Theories of Colored Bodies and Fits of Easy Reflection* (Cambridge: Cambridge University Press, 1993). Concerning science as public spectacle, Geoffrey V. Sutton, *Science for a Polite Society: Gender, Culture, and the Demonstration of Enlightenment* (Boulder, Colo.: Westview Press, 1995), and Simon Schaffer, "The Consuming Flame: Electrical Showmen and Tory Mystics in the World of Goods," *Consumption and the World of Goods*, ed. John Brewer and Roy Porter (London: Routledge, 1993).

90. Descriptions of these machines exist in several versions. Poleni prepared a list in Latin for use by Jacopo Facciolati for the *Fasti Gymnasii Patavini* (Padua: 1758), where the present instrument is listed on p. 423. See also Saladini and Pancino, *Il 'teatro' di fisica sperimentale*, 23, no. 49, based on an Italian version in BNM. On the major contributions of the Leiden shop of Samuel van Musschenbroek and his heirs Jan and Petrus, see Maurice Daumas, *Les instruments scientifiques aux dix-septième et dixhuitième siècles* (Paris: P.U.F., 1953), 325-27.

91. J. Facciolati *Fasti*, 426, G. A. Saladini and M. Pancino, *Il 'teatro' di fisica sperimentale*, 36, no. 105.

92. J. Facciolati *Fasti*, 418; G. A. Saladini and M. Pancino, *Il 'teatro' di fisica sperimentale*, 61, no. 189.

93. J. Facciolati *Fasti.*, 425; G. A. Saladini and M. Pancino, *Il 'teatro' di fisica sperimentale*, 74, no. 243. On electrical research in Italy in the first half of the eighteenth century, S. Ramazzotti and Luigi Briatore, "Appunti di storia della fisica: dalle calze di seta di Symner all'elettroforo di Volta," *Giornale di fisica* 15 (1974): 52-9; John L. Heilbron, *Electricity in the Seventeenth and Eighteenth Centuries* (Berkeley: University of California, 1979), 362-72.

94. J. Facciolati, *Fasti*, 418; G. A. Saladini and M. Pancino, *Il 'teatro' di fisica sperimentale*, 59, no. 179.

95. Compare 'sGravesande, *Physices*, 2, Bk. 2, ch. 5.

96. Facciolati, *Fasti*, 414. Compare 'sGravesande, *Physices*, 2, Bk. 4, chap. 5; Petrus van Musschenbroek, *Essai de physique . . . traduit du Hollandais*, 2 vols. (Leyden: Samuel Luchtmans, 1739), 2, chap. 7.

97. J. Facciolati, *Fasti*, 426.

98. J. Facciolati, *Fasti*, 413; G. A. Saladini and M. Pancino, *Il 'teatro' di fisica sperimentale*, 48, no. 145.

99. BNM, cod. ital. cl. IV: 592 (=5555), fol. 194, 21 April 1739: "Sono rimasto male in assetto per l'incomodo, cui non si può reggere, del luogo, per la folla, per il caldo, e per la gran fatica di dire e fare, e procurare assieme che l'uno e l'altro aggradevole e pulito riesca."

100. BNM, cod. ital. cl. IV: 592 (=5555), fol. 185, to the same, 18 March 1739: "L'operare, il lavorare in presenza degli altri, il dichiarare le cose con le lezioni, che nel mostrare si faranno agli uditori, i quali saranno nello stesso tempo spettatori dello speri-

mento, in una parola, la pratica mi aiutaranno a costituire quel sistema che mi sembri più adatto per ben servire."

101. Here and below, BNM, cod. Lat. cl. VIII: 153 (=2735), n.p.

102. On Bose, see the entry by John L. Heilbron in DSB 2 (1979), 324-25. On Banks, Albert Edward Musson and Eric Robinson, "Science and Industry in the Late Eigtheenth Century," *English Historical Review* 13 (1960): 233-34.

103. Margaret Rowbottom, "The Teaching of Experimental Philosophy in England, 1700-30," in *Actes du onzième congrès internationale d'histoire de la science, Varsovie, 1965* (Warsaw: Ossolineum, 1968), 4: 46-53; Willem Dirk Hackmann, "The Growth of Science in the Netherlands in the Seventeenth and Early Eighteenth Centuries," in Maurice P. Crosland, ed., *The Emergence of Science in Western Europe*, 89-110.

104. BNM, cod. ital. cl. IV: 592 (=5555), fol. 129: "Credo che in tal forma nascerebbe come da se stesso un ordine utile, chiaro, e scientifico, onde dappoi si potrebbe fare un corso in tutto e per tutto regolare, secondo un metodo esatto."

105. For a discussion of this problem in general, Ugo Baldini, "La scuola galileiana," in *Storia d'Italia, Annali 111. Scienza e tecnica* (Turin: Einaudi, 1980), 383-468.

106. B. Dooley, *Science, Politics and Society in Eighteenth Century Italy: The "Giornale de' letterati d'Italia" and Its World* (New York: Garland, 1991), chap. 5.

Chapter 6:
Saving the Jesuit's Skin

By the middle of the eighteenth century, the power of periodicals to engage the attention and galvanize the interest of readers of all sorts and from all backgrounds was well known. According to Cesare Beccaria, that made them as superior to books for some purposes:

> The majority of people view a book as they do a man who wants to come into their affairs and reform their entire families; they are terrified of having the entire structure of their ideas overturned. But a periodical paper, which presents itself to you as a friend who wants just to whisper a word or two in your ear, and that may suggest just a few useful truths not in bulk but one by one, and that may remove one or another error out of your mind almost without your noticing it, is much more welcome and much more heeded. The difference between an author and a reader usually insults our self-conceit, because most people do not believe themselves able to write a book; but everyone thinks he can write a periodical.[1]

For Beccaria, popular journalism was a particularly apt instrument for conveying messages of enlightenment.

And for conveying messages about science, the next best thing to popular journalism were the encyclopedic journals and magazines aimed at educated audiences. This is not to disparage more specialized journals such as the *Commentarii* of the Bolognese Istituto, which, like the *Philosophical Transactions* of the Royal Society, were uniquely devoted to science. But more readers had access to Angelo Calogerà's Venetian *Raccolta d'opuscoli scientifici e filologici*, the *Giornale de' letterati* of Ridolfino Venuti and collaborators in Rome, Giovanni Lami's *Novelle letterarie* in Florence, and Francesco Antonio Zaccaria's *Storia della letteratura italiana* in Modena and then Bassano, on which we will be focusing in this chapter.[2]

In spite of the mostly generic titles, a considerable portion of the contents of these encyclopedic journals and magazines was devoted to science. In Zaccaria's *Storia della letteratura italiana*, for instance, some seventy-five pages out of each five-hundred-page annual volume concerned the topics we have been defining as having belonged to the emerging disciplines connected with mathematics and natural knowledge. These journals responded to a growing market for scientific knowledge of all kinds. They fulfilled Galileo's promise of bringing science to a wider public. Science itself was changed as a result.

Of course, not all eighteenth-century encyclopedic journals were alike; nor did their purposes exactly coincide. Calogerà's *Raccolta d'opuscoli*, like Galileo Galilei himself, appealed to a larger public to gain support for new investigative methods and new forms of knowledge, against the supporters of more traditional methods. Zaccaria's *Storia letteraria*, on the other hand, appealed to a larger public to gain support for the Jesuits' role in science against the publicists of a rapidly forming Black Legend. In this curious reversal, a member of the order traditionally associated with the enemies of Galileo sought to use the same publicity tactics pioneered by Galileo. What is more, Zaccaria based his defense on what he conceived to be the order's long-time involvement in the very same scientific movement that Galileo had begun. According to Zaccaria, cultivating the most advanced forms of philosophical knowledge was a good thing; and the Jesuits had been a part of this.

Not everyone was persuaded. Almost as soon as the *Storia letteraria d'Italia* first appeared on the Italian publishing scene in 1750, it became embroiled in controversy. "It ought to be called the 'Storia letteraria of the Company of Jesus,'" wrote Paolo Maria Paciaudi to the Rimini naturalist Janus Plancus, intending no compliment; "because there is not a single page in it that does not praise some Jesuit."[3] Even a sympathizer like Scipione Maffei, learned amateur and literary conscience of northern Italy, did not expect it to amount to much.[4]

But nine years and fourteen volumes later, when the *Storia* finally stopped circulating its twelve hundred-odd copies to a readership that customary practices of sharing may have multiplied to nearly ten times that number, bringing the printer "incredible earnings" that continued through several offshoots, the journal had made a considerable contribution to the genre of literary and scientific publication.[5] By this time, the marketplace demanded information concerning the more innovative intellectual trends. Giving full coverage and encouragement to work by Giambattista Morgagni, Giovanni Targioni Tozzetti, Vincenzo Riccati, Ruggero Giuseppe Boscovich, and others, judging the best contributions and reinforcing the standards that the most innovative scientists set for themselves, Zaccaria's journal enriched the scientific culture of Italy between the age of Antonio Vallisneri and that of Galvani and Volta. Those who could provide such information were rewarded with popularity and esteem. Of this, the Jesuits needed all they could get, as the century wore on and the order's suppression loomed on the horizon.

According to eighteenth-century opinion, the Jesuits were not the only proponents of reactionary philosophical and scientific approaches and outmoded educational policies. Resentment still festered against Neapolitan ecclesiastics of all sorts, and not just the infamous Jesuit Giovanni Battista De Benedictis, who persecuted the atomists in the so-called "atheist trials," placing, according to Pietro Giannone, a "most heavy yoke . . . upon the shoulders of us Neapolitans."[6] The Roman Holy Office's recent condemnation of Francesco Algarotti's 1737 *Neutonianismo per le dame*, ostensibly because of its propagandistic tone, was widely viewed as the work of the Dominicans.[7] University of Padua naturalist

Antonio Vallisneri directed his scorn indiscriminately against "friars and . . . priests" of every sort, especially those "of Rome."[8] And still in this period, the Franciscans maintained their reputation as inveterate Scotists.[9]

But the Jesuits were held more particularly responsible than anyone else for the decline of science in Italy. Lami summed it up in these doggerels published in Florence:

> If ever you've heard a Jesuit talk
> At fallacious speeches you'll certainly balk
> Molina, it seems is all that he reads
> And in physics on none but Ptolemy he feeds
> . . .
> There in the corner, where he's not worth a pr- - -
> He shits on himself, and on the muses does stick
> All his satirical stench.[10]

And whatever they might do, the Jesuits were still made guilty by association with Galileo's persecutors of a hundred years before—even Zaccaria himself, to whom the Tuscan reformer Bernardo Tanucci referred by quoting the seventeenth-century poet Benedetto Menzini: "How very powerful are the Guelphs and those who 'stuck Galileo with a papal stinger.'"[11]

To vindicate the Jesuits from such accusations, perhaps no one in the mid-eighteenth century was better equipped than Zaccaria. Even before embarking on the journal, he began to distinguish himself by an unusually voluminous authorship of scholarly and polemical works, which in the end was to amount to over 150 titles in theology, patristics, ecclesiology, ethics, politics, and antiquities, including multiple editions.[12] And after succeeding Ludovico Antonio Muratori in the prestigious custodianship of the Este family library, he began directing a considerable entrepreneurial operation to distribute these works—as well as those of others like-minded—to centers all over the peninsula and beyond.[13]

During the course of his ten-year preaching apprenticeship in the cities of north-central Italy, while treading some of the same paths as the Jesuit missionary Paolo Segneri, he matched wits with Muratori, Daniele Concina, and Giovanni Lami in some of the great theological debates of the times. In each case, he tried to join his preaching activities with his theological ones by transforming polemics that began in Latin among a tiny number of experts into polemics in Italian among what he fully recognized would be an audience including the same assorted crowds he encountered in church. "A book . . . in the vernacular," he explained, "circulates among everyone."[14] At some point during these activities he must have become aware of the advantages of a periodical. "Apologies are read by few, if they are big volumes," he later remarked, "and if they are handbills, they are apt to get lost; this does not happen with articles of a journal, which are read by many and last for the whole duration of the work."[15] He credited political gazettes and handbills with having brought a knowledge of geography to everyone: "Even the common people, the second-hand dealer and barber know where are Petersburg, Memel, Stetino, Stockholm, and the Baltic Sea"; he no doubt hoped to do the same for the main propositions of modern theology.[16]

When he began the *Storia*, he continued the same polemics with renewed vigor. He supported the popular Marian piety that many critics had condemned as a relic of an outmoded Baroque spirituality.[17] It was no less essential, in the effort to reach the humblest believers, in his view, than was the probabilistic moral theology usually associated with his order, that critics viewed as permissive and lax.[18] He defended the wisdom and grandeur of Christ's apostles, and the present ecclesiastical hierarchy by extension, against suggestions, likewise reflecting upon the present, about the original ignorance and simplicity of both.[19]

Intending to undermine criticisms of the Jesuits' role in science, Zaccaria enlisted Jesuit mathematician Lionardo Ximenes to co-write the section on philosophy, mathematics, and natural knowledge for the first eight volumes of the *Storia*.[20] It was an inspired choice. Born in Trapani, Sicily, Ximenes had met Zaccaria for the first time in Rome while completing his theological instruction after a brief teaching tour of the Jesuit colleges of Florence and Siena. There he apparently came under the influence of Boscovich.[21] On completing his studies Ximenes accepted an invitation to go to Florence, where he published his famous treatise on sun dialing (*Del vecchio e nuovo gnomone*) and took a post as grand ducal mathematician. Soon he became involved in the two most ambitious land reclamation projects of the century: the drainage of the Pontine marshes around Rome and of the Maremma near Siena. Among the busiest Jesuit mathematicians, indeed, "the greatest Tuscan hydraulic engineer of the age," Ximenes could be depended upon to set an example of Jesuit science at its best.[22]

Together, Zaccaria and Ximenes directed their articles in these sections to providing unique information about a host of Jesuit practitioners who they claimed had been slighted or entirely ignored by their lay contemporaries—Francesco Maria Plata in Trapani, Niccolò Arrighetti in Siena, Antonio Lecchi in Milan, and Vincenzo Riccati in Treviso.[23] Most importantly, they provided perhaps the earliest simple account available in any modern language of Boscovich's matter theory.[24] Extracting his ideas on the subject from the treatise *De lumine* published in 1748, they explained how he conceived of the elements of matter to be not particles but true mathematical points, somewhat like the monads of Leibniz. They accounted for his insistence upon the necessity for a repellent force in nature, vaguely suggested by Isaac Newton, holding these points at a proper distance from one another, along with the attractive force that brought them together. Drawing upon direct experience mainly in chemistry, he believed only such a force could explain hardness and density and the lack of copenetration and transmutation between bodies under normal conditions. And they described his diagram of the law determining the interrelations between this force and the force of attraction, consisting of a curve showing the relative change in the two forces on one axis with changes in distance between particles on the other, so that repulsive forces increase infinitely as the distance between bodies becomes infinitely smaller. And after indicating at least a few of the broad claims he was to make for himself for this system, they presaged the enthusiasm that would greet his masterwork, the *Philosophiae naturalis theoria* ten years later. "Everything is full of ingenuity," they commented "and an admirable force of reasoning."[25]

If the *Storia* authors were to some extent guilty of the pro-Jesuit partiality Janus Plancus' correspondent attributed to them, their coverage of the fields of natural knowledge was by no means entirely one-sided. Not only because they covered as many Scolopians, Camaldolites, Franciscans, and Cistercians as they did Jesuits in these fields, and at least as many lay researchers, particularly in the field of medicine, where ecclesiastics rarely went. Also, they showed no reluctance to criticize work by Jesuits when incommensurate with the highest contemporary standards—as in the case of Arrigheti, who had asserted, among other things, that the moon did not gravitate toward the earth. "It cannot be said that there is nothing reprehensible in this theory," they observed, "for both the Cartesians and Newtonians and all the modern physicists agree that the moon gravitates forcefully toward the earth, like a stone or any other heavy body."[26]

Indeed, Zaccaria, in overseeing the production of those three-quarters or so of the journal that were devoted to the fields of natural knowledge and all other non-religious subjects, may well have intended to provide not so much an apology as an attractive, useful, complete, and for the most part dispassionate encyclopedic tool, like the analogous Jesuit *Mémoires de Trévoux*, to entice potential customers to buy into the more controversial part.[27] He certainly set out with every appearance of wishing to follow the example of the defunct encyclopedic Venetian *Giornale de' letterati d'Italia*, rightly viewed as the Italian forerunner of what had become one of the most commonplace literary genres of the time.[28] During the course of his preaching tour around Italy, he had occasion to meet Apostolo Zeno, the *Giornale*'s former director, who attended several of his Lenten sermons in Venice and may even have suggested the project. And he actually began to collect material while on his last preaching stop, in Florence, where he could observe the next successful encyclopedic journal in action, Giovanni Lami's *Novelle letterarie*, and take Ximenes on board.

And for publishing an encyclopedic journal, Zaccaria's varied background was in some ways ideal. Son of a prominent Florentine lawyer operating in Venice, he began his education in the Jesuit college there. The tale of his engaging in a famous theological disputation at age thirteen may be apocryphal, but already at twenty-three, after a novitiate in Vienna and a stint teaching rhetoric in Gorizia, the Austrian province sent him in 1737 as one of two candidates for theological instruction at the Collegio Romano. His horizons began to broaden as he befriended a future ecclesiastical historian and librarian (Pietro Lazzari), a physician (Giuseppe Benvenuti), as well as Ximenes, Boscovich, and perhaps Orazio Borgondio.[29]

Still lacking real expertise in any field except theology, Zaccaria acquired yet another field during his preaching tour of the cities of Italy—local history and antiquities. Wherever he went, he devoted most of his free time to visiting libraries and conversing with local scholars, with appreciable results. Besides Apostolo Zeno, a literary historian, and Giovanni Lami, who was not only a journalist but also a founding member of the Colombaria Society dedicated to ancient and medieval antiquities and publisher of collections of scholarly dissertations on those subjects, he also met Angelo Maria Querini, the bishop of Brescia and a skilled orientalist. In his *Excursus litterarii per Italiam ab anno 1742 ad annum 1752*, modeled on Jean Mabillon's *Iter italicum*, he recorded his journeys, including lists of monuments and codices previously neglected or badly described by others, occasionally even giving variant readings. And in so doing, he earned the almost unqualified praise of the Royal Society of Göttingen's *Göttingische Anzeigen von Gelehrten Sachen* that reviewed the book.[30]

What he lacked in expertise in the fields of natural knowledge Zaccaria made up for by his faith in self-improvement. In a letter to a young Brescia nobleman named Lorenzo Covi, he offered encouraging words concerning the possibility of being able to contribute something important in spite of the apparent abundance of brilliant minds. The mystique of authorship, he said, inevitably dissipated when modern productions were divided into their three large categories: transcriptions or editions of authoritative writings, compilations or collections of works written by others, and, finally, original works. All these categories invited new contributions, even the last; for no matter how many books were in existence, there still remained something new to say.[31] He gave the example of what he preferred to call "philosophy and mathematics"—his blanket term for the exact sciences. Here, no adequate complete course on philosophy yet existed—a complaint he and Ximenes were to repeat later in the *Storia*, with an incitement to his readers to try their hand at providing one; and almost all the major issues were still in dispute.[32] Never be too impressed by the so-called experts, he warned.

His contribution to the coverage of the fields of natural knowledge in the first eight volumes of the *Storia* was the masterpiece of Zaccaria's auto didacticism. He could not possibly claim a perfect understanding of mathematics, biology, botany, physics, chemistry, engineering, surgery, pharmacology, and medicine. But the remarks he wrote up with Ximenes' help showed a satisfactory comprehension of the material, and the remarks he made entirely on his own in biology and medicine occasionally included original observations. "Permit me, who have proposed to narrate the discoveries of others, to offer a suspicion of mine to the judgment of the scholars," he once said, introducing a theory of his own about the formation of the blood.[33] he was willing to defend himself when attacked for offering unauthorized medical views. "We cannot judge differently—not because we are, as [our opponent] gives us the honor of declaring us, simpletons and completely devoid of these matters and like parrots simply repeating what others say, but because the reasons he brings to bear to support his cause are prejudices and romances rather than arguments worthy of his learning, his rank and his age."[34]

Having chosen a book review format rather than substantive articles, and an annual rather than a monthly or quarterly publication schedule, it remained to Zaccaria and his collaborators, including Ximenes, to choose an appropriate editorial voice. Rather than the intimacy offered by the popular invented letter strategy used by the *Memorie per servire alla storia letteraria*, they offered levity. In the section on mathematics they proposed a "Problem concerning the lottery: Find a very easy way of winning any given amount of money in the ordinary game of the lottery." The answer: "Don't play."[35] They usually found that true stories offered irony enough, as in the case of a manifesto by one Valentino Roveda, hermit in Asti, who claimed to demonstrate the pointlessness of all geometry. "His learning is no less than his devotion. But it is so sublime that I confess I don't understand it at all. Who am I, daring to soar behind this great eagle?"[36] And on the same occasion, they said: "While all these geometers have been trying to raise the great edifice of Geometry and analysis, they probably did not notice that their whole building was tumbling down and collapsing at its foundations. But I believe it is my duty to warn them not to build so high as to find themselves one day in ruins."[37]

Most of all, they took pains to make readers comfortable with unfamiliar material by gradually immersing them in the field rather than plunging them headlong. Mathematics, in case anyone had not yet heard, was "noble and vast"; and in recent times conspicuous for its important contributions to physics.[38] Physics, on the other hand, included fashionable subjects like electricity, which—the strange scientific culture of the times required them to repeat—engaged even women.[39] Chemistry was especially necessary for the study of medicine.[40] And medicine, they explained, was a more respectable science than most people might think by reading Petrarch's denunciations of it, as long as it was based on the study of nature.[41] General criticisms of whole fields were occasionally in order, such as in geometry, where commentators on Euclid's *Elements* always "refried" the same old things.[42] And all fields in general sometimes deserved reproach for practicing the sorts of arguments "that even dogs would be ashamed of"—in spite of the abundant availability of works on the art of reasoning.[43]

For their readers' greater information, they provided simple laymen's summaries of the main issues behind complex debates. For example, before going into a series of works on anatomy they explained in lengthy detail how putrefaction and fermentation in the stomach produced the nourishing substance called chyle, part of which subsequently passed into the mesenteric veins and eventually on to mix with the blood in the heart.[44] And before going into a series of works on mechanics they explained the problem of the so-called "live forces" (*forze vive*). A little misleadingly, they told how Leibniz had once asked whether heaviness, magnetism, centrifugal force, and so forth, acting on bodies at rest could be measured the same way as the force possessed by a body because of its motion. And he had answered no; the two situations were different in kind and the forces present in the first situation should be called "dead" and those in the second "live." "Since the manifestoes explain it little and the printed books do not treat it sufficiently," they noted, "this question is therefore in the mouths of everyone but in the brains of few."[45] So into the brains of their readers they poured information about recent contributions by Vincenzo Riccati and, from the Bolognese Istituto, Francesco Maria Zanotti.

To introduce readers to a scientific way of thinking, the *Storia* authors showed exactly what sorts of everyday problems experimental method might solve:

> In the seawaters around the said city of Chioggia one often sees, especially in the summertime, a movement of many tiny flames. Signor [Giuseppe] Vianelli decided to apply his learned curiosity to finding out what these could be.[46]

Readers then discovered how to form and test a hypothesis:

So he collected some of this seawater and took it home; and after he passed it through a closely woven linen cloth, he began striking the water to see if it still sent out the little flames as before; but they no longer appeared. Instead, he saw a sparkling material on the linen.

Instruments penetrated where the naked eye could not:

> He examined the material under the microscope; and thereupon he discovered that it was the cause of the nocturnal light that had made such a graceful scene on the seawater, and that it was made up of nothing other than previously undiscovered insects.

The testing of nature yielded up more than the answer to the original question; it yielded the unexpected identification of an entirely new object.

Other experiment narratives showed the importance of serendipity, while emphasizing the interest in science among learned virtuosi belonging to the upper classes:

> While working on a chemical operation in July 1752, for use in certain experiments, after four months of work, [Prince Raimondo di Sangro] produced a certain material that he placed in some glass jars. Now, in the last days of the following November, when opening up one of the jars that contained a quarter-ounce minus seven grains of the said material, and bringing it close to a lighted candle, the material caught fire.

Exact measurements were an important component of an empirical discipline, so those formed part of the story; although there were still phenomena that had to be measured by their relation to common everyday activities. In the present instance, "It shot out a yellowish flame, so beautiful and bright that one could read and write by it."[47] Endless repetition was the key to success, as the accumulation of knowledge progressed by tiny increments and produced many frustrations before yielding the satisfaction of a real discovery:

> After six hours, he extinguished the flame by covering the jar; but the next morning he was unable to light it again. The same thing happened another time when he made a sort of candle out of the said substance, and it burned continuously from the last days of November to March 2, 1753, and beyond; but when it was put out by an accidental bump during the course of other experiments, the prince could not light it again, and the substance remained strangely inert and incapable of sending out any more flames.

In keeping with the Enlightenment conception of knowledge, the next step was to find a practical use: "In both cases, the prince observed that with all the burning, the substance neither changed nor diminished by a single grain. If the Signor Prince manages to popularize this discovery," the authors quipped, "the price of oil will collapse."

To be sure, this was not great science; but on other occasions, Zaccaria and Ximenes provided technical information that might be of more use to the experts. Accounts of the solar eclipses of 8 January and 19 June 1750, gave measurements from four different locations, and of the lunar one of 8 June 1751, from two locations. They described the transit of Mercury across the sun on 5 May 1753, as seen in Bologna.[48] They reproduced Ximenes' calculations for deriving the correct difference in longitude between Pisa and Livorno. They reproduced Giuseppe Veratti's timely and exhaustive account of the successful attempt by members of the Bolognese Istituto to repeat Benjamin Franklin's experiments on electricity in clouds.[49] For those proficient enough to try, they provided a varied repertoire of geometrical problems, from Joseph Antoine Chautard Du Clos' method for inscribing an enneagon inside a circle, sent in from Turin, to Valentino Roveda's method, sent in from Asti, of squaring the circle by starting with a right triangle whose smallest side equaled the radius in length and whose hypotenuse equaled one-fourth of the circumference.[50] And for those interested in new instruments, they reported on the new Boylian air pump acquired by the Noble College in Naples, and, in Crema, Domenico Crespi's creation of a pendulum- and spring-driven equation clock with internal adjustments, prefiguring the work of Ferdinand Berthaud, to lengthen the month of February automatically every leap year.[51]

As "the broadest and most exhaustive critical bibliography yet produced in Italy," the journal eventually achieved critical acclaim.[52] The Marquis de Cursay, financier of the *Journal étranger*, sought Zaccaria as his Italian correspondent, and French minister Pierre Guérin de Tencin sought him as his bibliographical advisor. Antoine Gachet d'Artigny excerpted the journal for his *Nouveaux mémoires d'histoire, de critique et de littérature*.[53] Scipione Maffei changed his first lukewarm assessment; and Apostolo Zeno recommended publication in two yearly volumes instead of one.[54] As a commercial venture, the *Storia* succeeded where other similar ventures failed. Venetian printer Orazio Poletti assumed the production costs himself, and the first and second printings of the first volume failed to keep up with demand.[55] Soon the demand far outran the capacities not only of the first printer but also of Bartolomeo Solani, the small-time local printer Zaccaria engaged upon moving his whole operation to Modena as the new librarian of the Estense. Accordingly, Giambattista and Giovanni Antonio Remondini of Bassano took it over, the fastest growing printing firm in northern Italy.[56]

While conveying natural knowledge to a growing readership, the *Storia* authors sought also to instill a kind of intellectual patriotism. They joined the campaign, pursued with the greatest fanfare by the *Giornale de' letterati d'Italia*, against their fellow-Italians' reliance upon transalpine cultural affairs at the expense of what went on at home.

> If ever one could write 'London' as the place of publication of an Italian book, oh how much respect would it not encounter in Italy? It adds so much luster to a work to be able to say that it is well-traveled. A book born in Florence could not possibly be any good. It is necessary to bring it from overseas. Oh miserable condition of Italians, once the masters and the lords of the world, and now considered by more than one nation as scholars and slaves.[57]

They recited the age-old complaint about a transalpine tendency to steal Italian discoveries.[58] And recognizing that transalpine scholars were likely to remain more informed about Italian things than the Italians were themselves as long as the likes of Louis Bourguet of Neuchâtel provided instruments such as the *Bibliothèque italique* with this very end in view, Zaccaria and Ximenes followed the *Giornale* in limiting their coverage to works written in Italy.[59] And in so doing, they claimed to promote not only "the advantage of letters" but also, "the honor of the nation."[60]

In their view, the most important item on the agenda of contemporary Italian science was the recovery and interpretation of the methodological paradigm of Galileo. Not that Galileo's influence on Italian science had ever abated since the single works first appeared. But the first complete edition had only recently been published in Italy.[61] And remnants of the outmoded Aristotelian and scholastic physics still appeared in textbooks. Bringing the discussion of Galileo out into the open forum of the press might demonstrate the value of Italian science as well as remove some prejudices. That the authors were Jesuits might surely signify a new attitude on the part of ecclesiastical authorities.

Apart from a few well-intentioned mistakes, the *Storia* authors' assessment of Galileo was powerful and timely. They wrongly credited him with having abandoned entirely the geometrical method of Eudoxian proportions in his mature work.[62] But they correctly explained his mathematization of physics. "Galileo was the first," they noted, "who left the philosophy of the ancients and introduced the new and different one prevailing in present times."[63] They defended his conclusions on motion, the planets, and the corruptibility of the heavens in the *Dialogue Concerning the Two Chief World Systems*. They referred to his authority when assessing works treating odors and colors as mechanical effects upon the senses rather than as accidents or qualities. And in his name, they threatened by a menacing array of air pumps, telescopes, and compasses anyone who tried to hold back the progress of empirical knowledge.[64]

Again, in their account of the exploits of Galileo's direct and indirect disciples, the *Storia* authors missed a few cues. They credited Domenico Guglielmini with the main principle of hydrodynamic measurement—namely, that velocity differs between the surface and lower regions in any river—discovered in fact by Galileo's disciple Benedetto Castelli a hundred years before.[65] They were perfectly correct, however, in implying that Guglielmini profited from the continuation and development of Galileo's empirical method throughout the previous century. The pioneering Florentine Accademia del Cimento, with Giovanni Alfonso Borelli, Francesco Redi, Lorenzo Magalotti, and the rest, fashioned new instruments for applying it.[66] Two of the members, Redi and Borelli, extended it to the life sciences, and Marcello Malpighi, by relying exclusively upon extensive comparative observations, followed in their steps. Paolo Boccone, Malpighi's Sicilian disciple, explored new ways of applying it to botany, subsequently followed by the Florentine Botanical Society and its members Giansebastiano Franchi, Pier Antonio Micheli and others.[67] It showed up again in the "great work" on human generation by another Malpighi disciple, Antonio Vallisneri, who improved upon Antonie Van Leeuwenhoek's theory about tiny preformed creatures in the semen.[68]

This interpretation of the Galileian tradition, of course, meant soft-pedaling the various discordant notes that we examined in chapter 3. It meant leaving the impression that non-Galileians like Giovanni Battista Riccioli, Athanasius Kircher, and Sebastiano Bartoli never contributed anything important—even though they were Jesuits. And even among the Galileians, it meant forgetting the battles between Borelli and Vincenzo Viviani in the Accademia del Cimento. It meant playing down the chemical philosophy of Lionardo Di Capua and the cabalism of Elia Astorini. It meant ignoring the scriptural literalism of Giambattista Hodierna and Pietro Mengoli. It meant ignoring the current of vitalism extending from Donato Rossetti in Galileo's time all the way up to eighteenth-century naturalist Francesco Maria Nigrisoli. So instead of regarding as a symptom of a fractured tradition the most recent attempt, by Neapolitan philosopher Costantino Grimaldi, to resuscitate Giambattista Della Porta's method of cryptology and resemblances and apply it to modern problems, they simply left it out altogether.[69]

And on the basis of the supposed unity of the Galileian tradition, and not its fragmentation, the *Storia* authors registered their approval of current work. Michele Genorini, physiology instructor at the Florentine Studium, for instance, was perfectly Malpighian—i.e., Galileian—even though he aimed to disprove Malpighi's theory of the conversion of chyle into blood through the grinding action of the lungs. Emulating Galileo-style thought experiments, including a bag full of large and small glass balls to show that larger particles like those of the blood would actually get ground up first, Genorini offered his own view that the conversion came about through the lungs' action of extracting and exhaling chyle acids. And Malpighian in their methodological approach were several recent experiments to test the curative potential of mercury, on which the *Storia* authors believed physician Giuseppe Maria Saverio Bertini of Florence showed far more expertise than fashionable "desktop physician" Hermann Boerhaave of Leyden.[70] Finally, they did not forget to salute the Galileian school's characteristic combination of science and the humanities, by giving Giambattista Morgagni's authoritative edition of the works of two physicians of antiquity, Aulus Cornelius Celsus and Quintus Serenus, two long reviews.[71]

Fully in the Galilean school's tradition of naturalistic description, they insisted, were the recently published *Reports on Some Travels Through Various Parts of Tuscany* by Giovanni Targioni Tozzetti. In volume four, Targioni Tozzetti provided the reader with explanations of everything geological in sight on a meandering trip along the hilly western margin of the grand duchy from Barga to Monterotondo. And squinting back through the mists of prehistoric times, he decided that horizontal cracks and fissures there visible could only be explained by the buildup of tension between land masses of different specific gravities compressed in different ways. Georges Buffon's suggestion in the *Histoire naturelle* that a settling of the ground at the base of the hills, Targioni Tozzetti pointed out, was only plausible for vertical features. Furthermore, drawing upon Italian research on hydrodynamics, he showed the tremendous power of river currents to modify landscape morphology and criticized Buffon's theory tracing the origin of valleys and gorges solely to the action of marine currents before the land emerged. Something would have had to make those violent currents move in such irregular ways, he reasoned; but in the period hypothesized by Buffon there would have been no large protruding land masses. He thus explained the weirs of the river Torriti based on an age-long process of erosion. And he added a powerful explanatory tool to the emerging science of paleontology.[72]

While recommending local products with the same fervor as their predecessors on the *Giornale de' letterati d'Italia*, the *Storia* authors nevertheless paid due attention to the European context of the books they studied; and for the greatest living example of the marriage between physics and mathematics, they turned again and again to Isaac Newton. True, they did not always faithfully portray Newton's position in contemporary physics. Perhaps to render the theories more palatable to many Italian readers who were likely to prefer the theories of Descartes, they claimed, a little simplistically, that the main disagreement between the two camps was over whether gravity was to be attributed to the effect of the vortex or to a property of matter.[73] However, in other resumés of Newton's greatest discoveries, they cited passages from the *Principia* to show how he drew the true consequences of Kepler's laws of planetary motion. They showed how he compared the elliptical orbits described in Kepler's first law to the motions bodies must make if they were circulating around each other with a centripetal force varying inversely as the square of the distances. "Kepler . . . , trying to use his magnetic theory to give a mechanical explanation for such motions," they informed readers, "did not notice that the description of that curve was the result of a universal gravity."[74] They accounted for Newton's explanation of irregularities in the sun's movements based on its position near the entire system's center of gravity.[75] And they outlined his discovery of the relation between the moon's gravity and the changing tides on earth.[76]

When measuring contemporary Italian contributions to astronomy, the *Storia* authors used Newton's work as a guide. They condemned anyone who denied the universality of the law of gravity.[77] And they blamed Bolognese astronomer Eustachio Manfredi's posthumous *Istituzioni astronomiche* (Bologna: 1749) for cutting short a discussion of the Newtonian system in order to prolong a useless and dated one comparing the heliocentric with the now-defunct geocentric hypothesis.[78] Let readers instead pay attention to Boscovich's effort to clarify Newton's own thoughts, they insisted. His gloss marvelously clarified Corollary 4 of Newton's Third Law of motion, establishing that common centers of gravities of bodies do not change their state of motion or rest by the actions of the bodies among themselves but remain at rest or move uniformly in a right line. It was an important precondition of Newton's idea in Book 3 proposition 11 that the center of the universe was at rest.[79] Likewise, Paolo Frisi added persuasive proof to Newton's outdated observations in the *Principia* (Book 3 proposition 19) concerning the earth's bulging about the equator because of the forces generated by rotation. The ensuing quarrel between Newtonian bulging-equator theorists like Pierre-Louis Maupertuis and Cartesian bulging-pole theorists like Jacques Cassini brought about expeditions to Lapland and to Ecuador to get accurate on-the-spot measurements. Frisi actually set the Newtonian theory on a more secure foundation than ever by correcting tiny mistakes in field measurements committed by the excessively zealous Maupertuis.[80]

Again, for the example of empirical knowledge at its best, the *Storia* authors turned to Newton on optics. They outlined the problem he solved, starting with previous theories claiming color to be an intrinsic quality of a body, and they recounted the famous experiments dissecting white light and then reassembling it from its basic component colors. They showed how Newton arrived at his theory that colors come from rays of various types characterized by constant regular grades of refractibility and by the ability to produce different effects on the optic nerve. Objects therefore appear to show one color rather than another, they explained, because of different capacities for absorbing and reflecting the various wavelike rays of corpuscles. They then explained how the Tuscan Jesuit Pier Maria Salomoni further developed these ideas by suggesting that the maximum and minimum angles of the refracted rays producing the rainbow could be established by the differential calculus rather than by the time-consuming "synthetic" method preferred by Newton in *Opticks* Book 1 proposition 9 and by Newton's followers Willem Jacob 'sGravesande, Petrus van Musschenbroek, and others.[81] They praised Salomoni's linguistic aids, explaining what "indigo" might look like and how it differed from "violet" or "purple." Italian readers of Francesco Algarotti's popular *Neutonianismo per le dame* and the recent translation of Henry Pemberton's *View of Sir Isaac Newton's Philosophy* now had a third guide to modern thought.[82]

However, the *Storia* authors were not so taken with Newton's optics as to be incapable of appreciating the debates surrounding them, both in Italy and in transalpine Europe. They left alone the experimental results themselves, impugned by Scipione Maffei and others unable to repeat them because of defective instruments or fixed preconceptions. Instead, following the Venetian anti-Newtonian Giovanni Rizzetti, they impugned Newton's interpretation of those results. To the question of just how bodies could act at a distance on rays of light, causing all, some or none of them to be reflected, Newton's agnostic response was no help. "To tell the truth," they said, "we do not know what disposition and texture of bodies causes them to reflect the original red color rather than violet, or any other."[83] Arrighetti's simple mechanical explanation seemed promising. Could light not literally strike the hard particles of bodies and either pass through the pores, producing translucency, or stop dead, producing opacity? They seemed to share Arrighetti's difficulties in accepting an emission theory concerning the particles of light, because of the way light traverses very wide spaces in a very short time.[84] They showed how he imagined instead a very tenuous fiery substance, equally distributed throughout all the spaces in all things, serving light in the same way that the air served sound. As in Robert Hooke's wave theory, a very gentle impulse was propagated very quickly and successively throughout.[85]

Likewise, the *Storia* authors objected to the Newtonians' extension of the notion of attraction beyond the strict limits set for it by Newton himself. As an example, they cited Andrea Bina's "Newtonian" theory of electricity.[86] Of the same work, they recounted the theory that sunspots were simply clouds of objects that had been drawn to the sun by the force of attraction, like feathers to a glass rod. Whoever believed this, they said, could apply the same theory to the Aurora Borealis. "Anything is possible," they quipped, "when reasoning with the aid of P. Bina's attraction and mechanicism."[87] The same went for John Keil's "Newtonian" theories of medicine, criticized by the Sienese physician Pietro Cornacchini for attributing the circulation of the blood and even the separation of the humors to an attractive force. The unnecessary extension of "the laws" of this force, they warned, was "fast becoming a ridiculous and very dangerous abuse."[88]

However, unlike many exponents of the so-called "Catholic Enlightenment," the *Storia* authors did not balk at this widespread use of the concept of an invisible attractive force merely for fear of mixing physics and metaphysics or even science and theology. Following in Galileo's footsteps, it is true, they did everything they could to get the imaginary ghosts out of the world machine. They shunned literal interpretations of Genesis in favor of a creation scheme incorporating the geological evidence offered by Vallisneri and Giacinto de' Tonti.[89] They interpreted witchcraft as a social ill proceeding from popular ignorance and not a spiritual ill, at least in most cases, based on sociological evidence offered by Girolamo Tartarotti, despite the ambiguous references in Deuteronomy.[90]

While engaging in these efforts to get imaginary ghosts out, they engaged in others aimed at making sure what they viewed as the real ones stayed in. This was, after all, a Jesuit publication. "The philosopher," they pointed out, "must concern himself with invisible beings no less than the theologian."[91] And in the investigation of invisible beings, philosophy could serve as the handmaid of theology far more than Galileo had ever dreamed. They encouraged efforts to use physics for studying levitation and other powers attributed to spirits, just when the Congregation on Rites under Benedict XIV was starting to use modern science for verifying miracles in the preliminary stages of the canonization process.[92] And they came down on the affirmative side in the great controversy over whether animals had a soul. This was so, they argued, because the Cartesian position that animals moved and performed all their operations by purely automatic and mechanical actions came far too close to suggesting that man, too, might be a sort of automaton. And Lorenzo Barbieri's suggestion that animals, though entirely mechanical, were nonetheless directed in their actions by divine Providence, was no way out, since it seemed to sneak a Cartesian principle of aspiritual lifelessness right back in. They thus endorsed the position of Lorenzo Magalotti, later shared by Antonio Genovesi, that the animal soul is simply an inferior variety, even though human effort seemed unable to decide. Apparently, at least a few features of the mechanical philosophy could be sacrificed where dogmatic truth was at stake.[93]

The extension of Newton's principle to more and more areas of reality furnished the *Storia* authors with an example of the negative effects of a marketplace upon science. Broad claims were the natural effect of the new social relations of science. "We ought to sympathize with our poor philosophers. They want to raise a hue and cry; they want to be applauded, and among most people this cannot be gained except by speaking about causes, and by putting together a huge machine of a system."[94] Yet, calling upon "deus ex machina" when all else failed, was unacceptable. Newton's name alone was no scientific argument, however well accredited. A new system explaining all phenomena could only be established over a long period of time and through the general agreement of most of the practitioners. And in spite of Newton's imposing achievement, no such system yet existed.

The *Storia* authors therefore made several recommendations. The first was caution—in other words, exactly the opposite of what they observed in field of electricity, where on the slenderness of evidence new systems and new theories were being erected every day.[95] The next recommendation was to avoid excessive hero-worship. Neither Newton nor Descartes ought to be invoked for inspiration on matters they had never studied. Between the two, safe from the fashions and the passions of the moment, stood a middle ground—a middle ground which Italian science, because of its traditions and because of its particular specialties, was ideally suited to not just to occupy but even to dominate. "How desirable it would be," they exclaimed, "with the French transported by the spirit of Cartesianism and the English by that of Newtonianism, and the other nations lined up on one side or the other, to see the Italians look upon these battles with indifference, and meanwhile dedicate their time to more useful researches than those concerned with the causes of things."[96] The final recommendation concerned the publication of information on science in the press. As a collaborative effort, science required the testing of theses by everyone involved; and submitting things to public judgment helped find truth.[97]

Most important of all, the authors appealed to an ever-wider audience for natural knowledge. In spite of the gains that had been made, there was still far too much indifference—especially among children, whose preparation in scientific fields was the only hope for future accomplishment. Little could be done about those whose modest means prevented them from advancing; but there was no excuse for the children of noble families. "The state of most of the noble youth in Italy is miserable."[98] Echoing Scipione Maffei's famous anti-noble diatribe in *The So-Called Science of Chivalry* (1710), they poured scorn on nobles who squandered their good fortune on gaming, the chase, and other more sinful pastimes, leaving to others not only the cultivation of intellectual pursuits but even the management of their own estates. Instead, they announced, abundant wealth carried with it the moral imperative to benefit society and to educate children to do the same.

Hard as they tried to please their readers and great as was their commercial success, Zaccaria and his collaborators nonetheless failed to silence their most vociferous critics. Just to name a few examples, Gian Vincenzo Patuzzi in Verona and Giovan Battista Macchi in Piadana complained about the *Storia's* treatment of rigorist morals and Church government.[99] Giuseppe Frova in Vercelli complained about its views concerning sacred images.[100] Concina complained that its scrappy style provoked the urgent self-defense of the people it had criticized in every field.[101] Giovanni Lami accused it of "distorted reasoning" and recommended it only to experts who could see through its "many errors."[102] Gaetano Fabbri impugned its coverage of medicine as amateurish and unbecoming ecclesiastics.[103] Janus Plancus thought it neglected important works—namely, his own.[104] And Gianlorenzo Berti in Lucca began a veritable anti-*Storia* industry, with a counter-journal called *Supplement* that ran for three volumes, followed by an *Anatomy of all the volumes*, excoriating it for excessive praise of sloppy research by Jesuits.[105] Even Tanucci was amazed by it all—although not unpleasantly. "So much anger, so much persecution. . . . But we want truth in the world and we will crucify."[106]

So vociferous were these critics, indeed, that they ended in dampening the enthusiasm of Zaccaria's own ecclesiastical superiors. Already in the third year of publication, Ignazio Visconti, the Jesuit General, demanded that volumes be sent for prepublication censorship all the way to Rome, limiting the considerable freedom in which Zaccaria had become accustomed to operating.[107] Soon he followed up this command with another. "Let me recommend once again that you observe greater moderation in your writing, and that you take care not to offend anyone, and even if provoked, observe the moderation necessary to defend yourself without offending others."[108] As the journal's reputation for making trouble continued to spread, Pope Clement XIII himself is reported to have exclaimed, "Oh that *Storia*; that *Storia letteraria!*"[109] And when no amount of warning and counseling seemed sufficient to blunt the work's polemical weapons, Visconti's successor, Lorenzo Ricci, finally pleaded with Zaccaria to suspend the journal in 1758 "to further the interests of the Order."[110] Thus, in spite of the continuous efforts of Francesco III d'Este, duke of Modena, to deflect the onslaught from Rome, Zaccaria finally gave up the prospect of a continuation of the journal under another title until four years later—although for the moment, other concerns were foremost in his mind.

For reasons quite unconnected with Zaccaria's activities, the whole order soon became caught up in a fatal struggle. Standard complaints about moral lassitude, philosophical immaturity, pedagogical obtuseness, and political meddling had by now finally been joined into a new conspiracy theory, far more potent than the many ones announced in the previous century.[111] It first emerged in Portugal, where João I's minister Sebastião João de Carvalho e Melo (later marquis de Pombal) sought to use the Jesuits as symbols of the opposition to his program of state modernization, economic reorganization, and educational reform. There were accusations about fomenting popular political discontent during the reconstruction following the Lisbon earthquake. There were charges of exploiting a privileged position in the colonies, especially in Paraguay, to compete commercially on unfair terms with state companies. Finally, allegations about complicity in the Távora-Aveiro assassination plot justified the expulsion of the order in 1758—from the schools, from the missions, and from the state.

In Italy, the Jesuits became the victims of a public sphere of philosophical discussion that demonized them as the sworn enemies of modern culture. The unprecedented pamphlet war included contributions by Italian Enlightenment philosophers like Tommaso Antonio Contin. "The poor read them," apostrophized another writer, reputedly Gioacchino Faranca, "and learn to cry vendetta for the blood that you [Jesuits] have sucked from their veins in your insatiable greed. The merchants read them and conceive mortal jealousy for their usurped commerce. The plebs read them, and horrified by your numerous excesses, begin to point their fingers at you, whistling loudly, as they do on Saturdays against the Jews."[112] Amid the general uproar, the conspiracy theory became a main component of the Enlightenment-led anticlerical movement of the 1760s. To stop the onslaught was far beyond even the considerable polemical gifts of Zaccaria, which he employed to the best of his abilities in a pamphlet counter-attack. And governments responded to the growing pressure for antiecclesiastical change by abolishing main mort, suppressing local offices of the Inquisition, dissolving ecclesiastical courts, confiscating and redistributing Church property, and, in Parma, Naples, Milan, and Venice, expelling the Jesuits.

Among the various conspiracies the Italian Enlightenment philosophers attributed to the Jesuits was the plot to destroy Italian science. Had not their power and resources been entirely directed to a continuous secret effort to undermine, attack, and suppress the most important discoveries, asked the Milanese philosopher-mathematician Paolo Frisi, not only of Galileo and Copernicus, but also Cavalieri, Huygens, Newton, Descartes, Gassendi, and the rest? Anyone could see that all new entrants were simply inducted into the imperatives of the order as members of a transsecular historical community united to the forces of darkness against the spread of knowledge. And in Italy, the Jesuits almost had their way. No wonder that Italy had to struggle to regain its position among the scientific cultures of Europe. "Those countries where the [Jesuit] institute reigned," he insisted, "[have] remain[ed] for a long time below the level of other places."[113] To Frisi, as to the other Italian Enlightenment philosophers, the suppression of the Order in 1773 was no simple diplomatic coup by the marquis de Pombal, in spite of all appearances.[114] It was the inevitable, logical, and just consequence of an epochal struggle for freedom from a pernicious episteme and a meddlesome Church.

But the substantial contribution of the *Storia* to a new intellectual category that more and more late eighteenth-century publications were calling "science" in our modern sense, could not be erased. For it may be true that more specialized journals, like the *Commentarii* of the Bolognese Istituto, did more for furthering actual discoveries; and single-sector publications like the Venetian *Giornale di medicina* and the Lucchese *Memorie sopra la fisica e la scienza naturale* did more for medicine and physics; just as technical publications like the Venetian *Giornale d'Italia* did more for the practical application of scientific knowledge to agriculture. Yet encyclopedic journals performed the necessary function of distinguishing the disciplines of natural knowledge and mathematics from sister-disciplines in the humanities and social sciences. The Bolognese Istituto was already devoted to "the sciences" from 1712. By the late eighteenth century, academies distinguished "the sciences" from "letters and arts" by their very names—for instance, the Accademia Patavina di scienze, lettere ed arti.

Apart from administering Jesuit propaganda, the *Storia* may well have contributed to the cultural ambience that prepared the late great eighteenth-century Italian scientific renaissance. True, many other causes collaborated to bring about this result. We will account for some of them in the next chapter. And in the forefront, helping to build a marketplace for scientific accomplishment, widening the pool of applicants, and bringing them to the attention of those in power, was the periodical press. In the last twenty years of the century, Andrea Rubbi's *Nuovo giornale letterario d'Italia,* one of the successors to the *Storia,* could rightly claim that Italian scientists, who had once "lost their primacy," had now finally taken over genuine leadership.[115]

In some ways, Zaccaria and others like him were victims of their own success. Because of their appeals, a marketplace developed that was increasingly open to recent trends. But markets have their own imperatives. Circuits of supply and demand, as we have seen throughout this book, are strongly influenced by the circumstances in which individual practitioners find themselves, by culture, by social and political contexts. In the case of the market for cultural commodities, the power of Church and state were never far away. Nevertheless, at least in the early modern period, deliberate manipulation rarely worked.

Try as they might, those who sought to direct the new demand for science toward the particular products they had on offer, rather than toward those offered by others the market deemed more qualified, were bound to fail. No matter what they said and wrote, the Jesuits had no chance against those belonging to the emerging profession of the university professor. The characteristics of modern intellectual culture emerged through the combined efforts of the practitioners of the various emerging disciplines, propagandists of various sorts, and a public hungry for healing, for improvement, for new myths, for leadership, and for novelty of whatever sort.

Notes

1. *Il Caffè*, ed. Sergio Romagnoli (Milano: Feltrinelli, 1960), 291: "Gli uomini di questo genere, cioè la maggior parte, considerano un libro come un uomo che volesse entrare ne' loro affari, e riformar tutta la loro famiglia; sono ributtati dal timore di rovesciar tutto l'edificio delle loro idee; e gli uomini invischiati, per dir così, nell'abitudine soffrono nel doverne esser tratti. Ma un foglio periodico, che ti si presenta come un amico che vuol quasi dirti una sola parola all'orecchio, e che or l'una or l'altra delle utili verità ti suggerisce non in massa, ma in dettaglio, e che or l'uno o l'altro errore della mente ti toglie quasi senza che te ne avveda, è per lo più il più ben accetto, il più ascoltato. La distanza che passa tra l'autore di un libro e chi lo legge mortifica per lo più il nostro amor proprio, poichè il maggior numero non si crede capace di fare un libro; ma per un foglio periodico ognuno si crede abilità sufficiente, essendo poi sempre la mole, e il numero i principali motori della stima volgare."

2. Giuseppe Ricuperati, "Giornali nell'Italia dell' 'ancien regime'," *La stampa italiana dal Cinqecento all'Ottocento,* ed. Valerio Castronovo and Nicola Tranfaglia, 2nd ed. (Bari: Laterza, 1986), 251. In addition, Idem, "Politica, cultura e religione nei giornali italiani del Settecento," *Cattolicesimo e lumi nel Settecento italiano,* ed. Mario Rosa (Rome: Herder, 1981), 65; Marino Berengo, "Introduzione," *Giornali veneziani del Settecento,* ed. Marino Berengo (Milan: Feltrinelli, 1962), xvi-xvii; and, especially, Simonetta Santucci, Martino Capucci and Carolina Gasparini, "Storia della letteratura italiana," in *La biblioteca periodica. Repertorio dei giornali letterari del Sei-Settecento in Emilia e in Romagna,* ed. Martino Capucci, Renzo Cremante and Giovanna Gronda, 2 vols. (Bologna: Il Mulino, 1985-87), 2: 31-222.

3. Quoted in Maria D. Collina, *Il carteggio letterario di uno scienziato del Settecento* (Janus Plancus) (Florence: Olschki, 1957), 80, letter dated Ravenna, 31 January 1750.

4. Scipione Maffei, *Epistolario*, ed. Celestino Garibotto, 2 vols. (Milan: Giuffré, 1955), 2: 1273, letter to Jacopo Maria Paitoni, dated 28 May 1750.

5. Circulation figures are from a letter by Zaccaria dated 6 June, 1752, quoted in Mario Infelise, "Gesuiti e giurisdizionalisti nella pubblicistica veneziana di metà Sette-

cento," Mario Zanardi, ed., *I gesuiti e Venezia: momenti e problemi di storia veneziana della Compagnia di Gesù: atti del Convegno di studi, Venezia, 2-5 ottobre 1990* (Padova: Gregoriana, 1994).

6. Giannone got his revenge in the *History of the Kingdom of Naples* (Naples: 1723), where I quote from Bk. 40, chap. 4. In addition, Luciano Osbat, *L'Inquisizione a Napoli. Il processo agli ateisti, 1688-97* (Rome: Edizioni di Storia e Letteratura, 1974).

7. I compared the varying accounts of this event in Vincenzo Ferrone, *Scienza natura religione. Mondo newtoniano e cultura italiana nel primo Settecento* (Naples: Jovene, 1982), 35 with Paolo Casini, "Le Newtonianisme en Italie," *Dix-huitième siècle* 10 (1978): 98, with the more fanciful one of Mauro De Zan, "La messa all'Indice del Newtonianismo per le dame di Francesco Algarotti," *Scienza e letteratura nella cultura italiana del Settecento*, ed. Renzo Cremante and Walter Tega (Bologna: Il Mulino, 1984), 133-47; but the best description of the opposition appears to come from Algarotti's correspondent Eustachio Manfredi, who cites the work's excessive "liberty" and use of "French expressions." In Algarotti, *Opere inedite*, 8 vols. (Venice: 1796), 1: 139, letter dated 11 August 1738.

8. St. Petersburg, Soltykov Library, Angelo Calogerà correspondence, consulted in the microfilmotheque of the Fondazione Giorgio Cini in Venice, vol. 29, dated 5 December 1727, fol. 10r.

9. Zaccaria ridicules them them in *Storia letteraria d'Italia* [=SLI] 2 (1751): 151.

10. [Giovanni Lami], *I pifferi di montagna, che andarono per suonare e furono suonati* (Leyden [but Florence]: 1738], 7, in the inimitable translation of Eric Cochrane, *Florence in the Forgotten Centuries* (Chicago: University of Chicago Press, 1973), 385-86.

11. Bernardo Tanucci, *Epistolario*, 10 vols. (Rome: Edizioni di storia e letteratura, 1980-), vol. 3, ed. Anna Vittoria Migliorini (1982), 79, letter dated 24 April 1753.

12. All are listed in Carlos Sommervogel, *Bibliothèque de la Compagnie de Jésus*, 10 vols. (Bruxelles: O. Schepens; Paris: A. Picard, 1890-1909), vol. 7, cols. 1381-1435. The standard works on Zaccaria are Donato Scioscioli, *La vita e le opere di Francesco Antonio Zaccaria, erudito del secolo XVIII* (Brescia: Vannini, [1922]), which is updated in several articles by Enrico Rosa, "Gli scritti e il carteggio di F. A. Zaccaria in un archivio della Guipuzcoa," *Civiltà cattolica* 80 no. 4 (1929): 118-30; "La vita e le opere di Francesco Antonio Zaccaria," Ibid. 81 no. 1 (1930): 339-51; "Nuovi documenti sulla vita e le opere di F. A. Zaccaria," Ibid., 509-17; "Pubblicazioni e tribolazioni del p. F. A. Zaccaria," Ibid., no. 3, 27-40, 121-30.

13. These aspects of his career are explored in Infelise, "Gesuiti e giurisdizionalisti"; and Luigi Balsamo, "Editoria e biblioteche della seconda metà del Settecento negli stati Estensi," *Reggio e i Territori Estensi dall'Antico Regime all'età Napoleonica*, ed. Marino Berengo and Sergio Romagnoli, 2 vols. (Parma: Pratiche Editrice, 1979), 2: 505-32. Consider also Franco Venturi, *Settecento riformatore*, vol. 2: *La chiesa e la repubblica dentro i loro limiti, 1758-1774* (Turin: Einaudi, 1976), 22; Émile Appolis, *Entre jansénistes et zelanti. Le "Tiers parti" catholique au XVIIIe siècle* (Paris: J. Picard,1960), 570; Giuseppe Pignatelli, "Le origini settecenteschi del Cattolicesimo reazionario. La polemica antigiansenista del Giornale ecclesiastico di Roma," *Studi storici* 11 (1970): 759n.

14. SLI, 12 (1758): 310. The comment concerned Muratori's *Della regolata divozione*. Some of Zaccaria's treatises in this period were *Lettere al signor Antonio Lampridio intorno al suo libro nuovamente pubblicato, "De superstitione vitanda"* (Palermo: 1741); *Lettere . . . sul libro: "De eruditione apostolorum"* (Venice: 1741); *Osservazioni sopra i primi cinque capitoli dell' Esame teologico* (Bastia: 1745).

15. From a letter dated 6 June 1752, cited in Infelise, "Gesuiti e giurisdizionalisti."

16. SLI 13 (1758): 212.

17. SLI 5 (1753): 430-44; 12 (1758): 310-12. I refer to Zaccaria's "Dissertatio prolegomena" to Alfonso Maria de' Liguori's *Theologia moralis*, 3 vols. (Rome: 1757). This issue is examined by Jean Delumeau, "S. Alfonso dottor della fiducia," *Alfonso M. De Liguori e la società civile del suo tempo. Atti del Convegno internazionale per il Bicentenario della morte del santo, 1787-1987*, ed. Pompeo Giannantonio, 2 vols. (Florence: Olschki, 1990), 1: 205-21 and Giorgio Petrocchi, "Sant'Alfonso scrittore mariano," Ibid., 2: 445-61. The wider context to Muratori's views has most recently been examined by Claudio Donati, "Dalla 'regolata devozione' al 'giuseppinismo,'" *Cattolicesimo e lumi nel Settecento italiano*, ed. Mario Rosa (Rome: Herder, 1981), 77-98; but also see Pietro Stella, "Preludi culturali e pastorali alla Regolata divozion de' cristiani," *Ludovico Antonio Muratori e la cultura contemporanea. Atti del convegno internazionale di studi muratoriani, Modena, 1972* (Florence: Olschki, 1972), 241-70.

18. SLI 1 (1750): 49-55; 5 (1754): 146; 12 (1758): 325-42. The standard study of Concina is still Alberto Vecchi, *Correnti religiose nel Sei-Settecento Veneto* (Venice-Rome: 1962), 307-400; for present purposes, 375-82

19. SLI 1 (1750): 41-2; 4 (1753): 404-22. The authority on Lami is still Eric Cochrane, *Florence*, Book 5; for present purposes, 338.

20. A note in the *Saggio critico della corrente letteratura straniera antica e moderna* 2 pt. 2 (1758): 316-67 clarifies this point: "Ora è da avvertire che incominciando dal tomo IX l'opera [*Storia letteraria d'Italia*] è di due altri autori, cioè, del p. Domenico Troili e del p. Gioacchino Gabardi. Il primo lavora i capi che alla filosofia, alle matematiche e alla medicina appartengono (benchè nel tomo IX il numero 7 del capo V del primo libro sino alla fine del capo sia d'altra mano, cioè del primario autore di quest'opera); l'altro i capi delle lingue, della poesia, dell'eloquenza e qualche altro, come nel t. IX il capo IX e nel t. X il capo della Storia profana. Tutti gli altri capi sono del primario autore; il che si avverte acciocchè ognuno sappia cui debba gli estratti delle sue opere. Per altro anche nel t. VIII il p. Troili ebbe qualche mano, e più negli altri ebbela il dottor p. Lionardo Ximenes, del quale benchè non tutti, son tuttavia parecchi estratti o di filosofia o di matematica, e quello massimamente pel quale i pp. Frisio e Bina han fatto tanto rumore." The account seems exact, as Zaccaria forewarned readers about the advent of a collaborative work already in the preface to SLI 5 (1754): "Non è da temere, che il diverso stile ostacolo sia ad aver nell'avvenire, quando che sia, compagni all'opera." Federico Sanvitale is mentioned as one of the "correspondents" in SLI 2 (1751): xii.

21. Information on Ximenes is from Luigi Palcani, "Elogio di Leonardo Ximenes," *Le prose italiane di Luigi Palcani* (Milan: Giovanni Silvestri, 1817), 7-34; Luigi Brenna, "Elogio del signor abate Leonardo Ximenes," *Giornale de' letterati* (Pisa) 64 (1786): 91-141; and Sommervogel, *Bibliothèque de la Compagnie de Jésus*, vol. 7, cols. 1341-57. Where his path crossed with Boscovich's is covered in Gino Arrighi, "Quarantaquattro lettere inedite di G. De la Lande, Ruggiero Giuseppe Boscovich, e Leonardo Ximenes," *La provincia di Lucca* 5 (1965); and somewhat superficially by the latest Boscovich biographer, Germano Paoli, *Ruggiero Giuseppe Boscovich nella scienza e nella storia del Settecento* (Rome: Accademia nazionale detta dei XL, 1988), chaps. 7 and 31. Concerning Borgondio, Paolo Casini, "Orazio Borgondio," in *Dizionario biografico degli Italiani* 12 (1970): 779. Borgondio explained his method in a letter to Antonio Vallisneri in Venice, Biblioteca Nazionale Marciana, cod. it. 148 (=6685), c. 11, 24 January 1716: "L'osservazione e sperienze saranno sempre il fondamento insieme e il contrassegno della vera fisica, appunto come l'osservazioni celesti sono l'appoggio insieme, e l'indizio della sussistenza nei calcoli astronomici."

22. The Tuscan projects are examined by Danilo Barsanti and Leonardo Rombai, *La "Guerra delle acque" in Toscana: storia delle bonifiche dai Medici alla riforma agraria* (Firenze : Medicea, 1986), from which I quote p. 14. The Roman one is surveyed in Hanns Gross, *Rome in the Age of Enlightenment* (Cambridge: Cambridge University

Press, 1990), 172-73. Ximenes' reports were collected in *Raccolta delle perizie ed opuscoli idraulici del Signor Abate Leonardo Ximenes*, 2 vols. (Florence: 1785-86).

23. SLI 10 (1757): 142-43 (Benvenuti); 2 (1751): 156-59, 3 (1752): 262 (Plata); 3 (1752): 268; 6 (1754): 130; 8 (1755): 69 (Arrighetti); 7 (1755): 131; 8 (1755): 46 (Lecchi); 7 (1755): 589 (Caraccilo); 8 (1755): 476 (Mangini); SLI 3 (1752): 245 (Riccati).

24. SLI 1 (1750): 128.

25. SLI 1 (1750):133. In their assessment of the importance of the work, they were at least as enthusiastic as Ivica Martinovic, "Boscovich's 'model of the atom' from 1748," *Bicentennial Commemoration of Ruggiero Giuseppe Boscovich, Milan, September, 15-18, 1987. Proceedings*, ed. M. Bossi and Pasquale Tucci (Milan: Unicopoli, 1988), 203-14. And they were on the mark as far as Boscovich's later claims were concerned, at least according to their analysis by Paolo Casini, "Ruggiero Giuseppe Boscovich," DBI 13 (1971): 225-26.

26. SLI 3 (1752): 271.

27. I comment on the *Mémoires de Trévoux* in B. Dooley, "From Literary Criticism to Systems Theory in Early Modern Journalism History," *Journal of the History of Ideas* 51 (1990): 482.

28. All references to that journal are based on my *Science, Politics and Society in Eighteenth-Century Italy. The "Giornale de' letterati d'Italia" and its World* (N.Y.: Garland, 1991).

29. His early friendships are mentioned in Scioscioli, *La vita e le opere di Francesco Antonio Zaccaria*, 13.

30. This work and the other major one of this period, namely, the *Bibliotheca Pistoriensis a Francisco Antonio Zacharia . . . descripta* (Turin: 1752), were reviewed in *Göttingische Anzeigen* (1755): 1368 and 1425. Muratori's text is in *Anecdota, quae ex Ambrosiana bibliothecae codicibus nunc primum eruit*, 4 vols. (Milan: Malatesta, 1697-1713), 2: 212ff. Mario Rosa evaluates the productions of these years in "Le 'vaste ed infeconde memorie degli eruditi': momenti della erudizione storica in Italian nella seconda metà del Settecento," *Erudizione e storiografia nel Veneto di Giambattista Verci*, ed. Piero Del Negro (Treviso: Ateneo, 1988), 19-23. Gori included work by Zaccaria in his *Symbolae litterarie*, 10 vols. (Florence: 1748-53), 4: 143-75.

31. "Lettera del Padre Francesco Antonio Zaccaria al sig. Lorenzo Covi cavaliere Bresciano sopra gli studi che da lui desidera intrapresi," *Raccolta d'opuscoli scientifici e filologici* 41 (1749): 89.

32. Ibid., 90: "Crederebbesi mai, che dove nella filosofia, nelle matematiche vantano i loro professori nuove terre per così dire discoperte, e nuovi mari, pur non avessimo un tolerabile corso di filosofia ed un pieno e sicuro trattato di matematica, che pressochè in ogni fisica question di qualche conto sperienze dovessimo vedere opposte a sperienze? Che in assai punti pro e contro recassersi dimostrazioni matematiche a gran meraviglia di chi penetra la forza e l'uso di questo termine 'dimostrazione' in fatto di matematica?" I compared SLI 8 (1755): 59.

33. SLI 3 (1752): 211.

34. SLI 3 (1752): 223.

35. SLI 3 (1752): 237.

36. SLI 3 (1752): 243.

37. SLI 3 (1752): 242.

38. SLI 1 (1750): 113. SLI 5 (1754): 70: "Non si corruccino i filosofi, se prima di parlare della lor facoltà discorriamo della matematica. Basti per ogni ragione sapersi, quanto alla buona fisica necessarie sieno le nozioni geometriche, e cento altre cose, le quali dalla sola matematica si possan prendere."

39. SLI 3 (1752): 258.

40. SLI 5 (1754): 151: "Se coll'aiuto del fuoco e delle ritorte non venisse a discoprire, quale, e quanta parte di sali, d'olii, d'acidi o d'alcaliche particelle è racchiusa ne' corpi, che a noi in varie maniere adoperati servono di medicina, come mai se ne potrebbono prescrivere le giuste dosi?"

41. SLI 6 (1754): 166: "Celebre è il detto di Francesco Petrarca, che non pure niente siavi a sperare da' medici, ma sì molto a temere. . . . Ma troppo esagerato è un tal sentimento. Perciocchè è veramente la medicina un arte di congetture, ma tuttavia ha ella i suoi sodi principii, da' quali un uomo d'ingegno e di sapere può utilissime conseguenze trarre a particolari bisogni degli uomini. Sopra ogni altra cosa dee un valoroso medico studiare la natura."

42. SLI 7 (1755): 128.

43. SLI 7 (1755): 144.

44. SLI 3 (1752): 205.

45. SLI 5 (1754): 71. More light has been cast on this quarrel by Thomas L. Hankins, "Eighteenth-Century Attempts to Resolve the Vis viva controversy," *Isis* 56 (1965): 281-97; Kathleen Okruhlik, "Ghosts in the World Machine: A Taxonomy of Leibnizian Forces," *Change and Progress in Modern Science. Papers related to and arising from the Fourth International Conference on History and Philosophy of Science, Blacksburg, VA, November 1982*, ed. Joseph C. Pitt (Boston: D. Reidel, 1985), 85-106.

46. SLI 2 (1751): 165. The information was from Vianelli's *Nuove scoperte intorno le luci notturne dell'acqua marina spettanti alla naturale storia* (Venice: 1749)

47. SLI 7 (1755): 200. The work in question was *Lettere del Sig. Raimondo di Sangro Principe di S. Severo di Napoli, sopra alcune scoperte chimiche* (Florence: 1754)

48. SLI 8 (1755): 477; 2 (1751): 502-12; 3 (1752): 651-56.

49. SLI 6 (1754): 686-94; 2 (1751): 512.

50. SLI 6 (1754): 684, 670.

51. SLI 7 (1755): 584, 589. Crespi's work is not mentioned in the standard repertory, Granville Hugh Baillie, *Watchmakers and Clockmakers of the World* (London: N. A. G., 1963), or in the DBI. I compared Ferdinand Berthoud, *Histoire de la mesure du temps*, 2 vols. (Paris: Imprimerie de la République, 1802), I: 188. David S. Landes discusses the equation clock as a technological feat, in *Revolution in Time* (Cambridge University Press, 1983), 123.

52. The comment is Vecchi's, in *Correnti religiose*, 375.

53. All this is in Scioscioli, "La vita e le opere di Francesco Antonio Zaccaria," 58.

54. Zeno's view is recorded in SLI 2 (1751): vii. Maffei's is in Scipione Maffei, *Epistolario*, 2: 1311, 24 May 1751, to Zaccaria: "Sappia che ammiro e lodo con tutti gli amici l'opera sua; e di quanto spetta a me, le rendo distinte grazie. Proseguisca pure, e procuri d'ottenere di non attender ad altro."

55. Infelise, "Gesuiti e giurisdizionalisti," quoting a letter of 6 June 1752, to Domenico Turano in Rome: "Il negozio è sicuro. Il libraio alle cui spese sinora si è stampata ha fatto incredibile guadagno, appena bastano le 1200 copie ch'egli ne ha tirate: è passata l'opera oltre monti ed è stata tradotta a Ginevra in franzese. Vuol dire che continuandosi e stampandosi a spese nostre il guadagno è certo."

56. For the role of the Remondini I refer to Mario Infelise, *L'editoria veneziana nel Settecento* (Milan: Angeli, 1989), 281-83. Zeno's endorsement is mentioned in SLI 2, preface.

57. SLI 5 (1754): 113.

58. SLI 8 (1755): 5: "Dall'Italia ha preso molto la letteratura straniera, e quanto indegni del nome italiano coloro [sono] i quali tutto prezzano fuor solamente le cose nostre." Maffei made a similar comment in *Osservazioni letterari* 1 (1737): xix-xx.

59. Francesca Bianca Crucitti Ullrich, *La "Bibliothèque italique": cultura "italianisante" e giornalismo letterario* (Milan: Ricciardi, 1974).

60. SLI 10 (1757): Preface.
61. Ferrone, *Scienza, natura religione*, 136n discusses the *Opere di Galileo Galilei* (Padova: Stamperia del Seminario, 1744), ed. Giuseppe Toaldo.
62. SLI 3 (1752): 234. I compared Stillman Drake, *Galileo at Work* (Chicago: University of Chicago Press, 1978), 58.
63. SLI 7 (1755): 146.
64. SLI 1 (1750): 101: "Qualcuno potrebbe . . . muovergli contro tutti i moderni coltivatori della fisica naturale, o sia della buona filosofia, e non so come volesse passarla, quando questi contro di lui rivolgessero e compassi e macchine pneumatiche, e telescopi, ed altri innumerabili strumenti loro." In addition, SLI 8 (1755): 58; 2 (1751): 152-53.
65. SLI 6 (1754): 100. Castelli's work was *Della misura dell'acque correnti* (1628).
66. SLI 1 (1750): 122.
67. SLI 1 (1750): 108.
68. SLI 8 (1755): 72.
69. Grimaldi's role is analyzed by Vincenzo Ferrone, *I profeti dell'Illuminismo. Le metamorfosi della ragione nel tardo Settecento italiano* (Bari: Laterza, 1989), 46. Other material in this paragraph is from the articles by Marco Ferrari and Paolo Galluzzi in *Scienze, credenze occulte, livelli di cultura. Convegno internazionale di studi. Firenze 26-30 giugno, 1980* (Florence: Olschki, 1982), 21-30, 31-62; Marta Cavazza, "Introduzione" and Gabriele Baroncini, "L'Arithmetica realis di Pietro Mengoli," in Baroncini and Cavazza, eds., *La corrispondenza di Pietro Mengoli* (Florence: Olschki, 1986), 1-22, 155-88; Paolo Galluzzi, "Il dibattito scientifico in Toscana, 1666-86," in *Nicola Stenone e la scienza toscana alla fine del Seicento, Convegno, Firenze 23 novembre- 6 dicembre, 1986* (Florence: Biblioteca Medicea Laurenziana, 1986), 113-30; Idem, "L'Accademia del Cimento: 'gusti' del principe, filosofia e ideologia dell'esperimento," *Quaderni storici* 16 (1981): 789-843; Maurizio Torrini, "Uno scritto sconosciuto di Lionardo di Capua in difesa dell'arte chimica," *Bollettino del Centro di Studi Vichiani* 4 (1974): 126-39; Eugenio Garin, *Dal Rinascimento all'Illuminismo* (Pisa: Nistri-Lischi, 1970), 135-44; Walter Bernardi, *Le metafisiche dell'embrione: scienze della vita e filosofia da Malpighi a Spallanzani, 1672-1793* (Florence: Olschki, 1986), 68-70, 112-19, and, in general, Paolo Galluzzi, Maurizio Torrini, Ugo Baldini, Elvezia De Angeli, Luigi Belloni, in Gino Arrighi, ed., *La scuola galileiana: prospettive di ricerca, Atti del Convegno di S. M. Ligure, 1978* (Florence: La Nuova Italia, 1979); and Baldini in *Storia d'Italia, Annali 3: Scienza e tecnica nella cultura e nella società dal Rinascimento a oggi,* Gianni Micheli, ed., (Turin: Einaudi, 1980), 383-468.
70. SLI 5 (1754): 181: "Nell'esaminare le materie mediche gioverà sempre oltremodo lo star lontani dalle ipotesi e l'accostarsi il più che possible sia alla sicurissima via delle sensate e giudiziose sperienze." SLI 1 (1750): 100-01: "il Boerhaave era più valente medico a tavolino che per esperienza di molte cure."
71. SLI 2 (1751): 132ff. Zaccaria referred to the following work of Morgagni: *Jo: Baptistae Morgagni in A. Cornelium Celsum et Quintum Serenum Sammonicum epistolae decem, quarum sex nunc primum prodeunt* (Padua: Comino, 1750). Some of the correspondence on which this work was based appeared in Quintus Serenus Sammonicus, *De medicina praecepta saluberrima* (Padua: Comino, 1722) and Aulus Cornelius Celsus, *De medicina libri octo* (Padua: Comino, 1722). The works are discussed in detail by Dante Nardo, "Scienza e filologia nel primo Settecento padovano. Gli studi classici di Giambattista Morgagni, Giovanni Poleni, Giulio Pontedera e Leone Targa," *Quaderni per la storia dell'Università di Padova* 14 (1981): 1-40.
72. SLI 5 (1754): 127. I compared Francesco Rodolico, "Giovanni Targioni Tozzetti," *Dictionary of Scientific Biography* 8 (1970): 257-58.
73. SLI 3 (1752): 271: "La dissensione tra Cartesiani e Newtoniani versa solamente in questo, che i primi questa gravità vogliono che sia un effetto del Vortice, ed i secondi

vogliono che sia una legge primaria della natura. Ma gli uni e gli altri accordano la gravità."

74. SLI 2 (1751): 139: "Il sig. Newton fu, che meglio penetrando il meccanismo celeste, dimostrò generalmente che, se un corpo qualunque graviti verso un centro per modo, che tal gravità sia in ragion reciproca duplicata delle distanze, e sia spinto con qualunque velocità, e con qualunque direzione (purchè non passi pel centro delle forze) esso sarà obbligato a descrivere una delle sezioni coniche, cioè, o una elisse o una parabola o un'iperbola, o un cerchio, considerando il cerchio come una sezione del cono. Da questa generalità venne il signor Newton a determinare per una costruzione geometrica la spezie dell'orbita medesima, distinguendo quali sieno que' casi, ne' quali il corpo sarà astretto a descrivere un'elisse, una parabola, un iperbola."

75. SLI 5 (1754): 82: "I Newtoniani hanno col loro maestro stabilito che il centro delle rivoluzioni de' primarii sia il centro comune di gravità de' primarii e del sole. Ma superando il sole di gran lunga nella sua massa le masse di tutti i pianeti uniti insieme, ne viene, che questo centro comune di gravità non è molto lungi dal sole medesimo. Indi è, che il sole medesimo diviene come un pianeta, il qual si rivolge intorno al centro comune di gravità; e siccome questo centro, che dipende dalle posizioni di tutti i corpi mondani sempre varianti, patisce una gran varietà, così non v'è orbita più irregolare dell'orbita, benchè piccolissima del sole."

76. SLI 7 (1755): 156.

77. SLI 3 (1752): 327: "Questo è un modo di ragionar di mercato vecchio. Questi argomenti e dicerie popolari non hanno luogo presso agli uomini dotti. . . . Oggi non vi è filosofo, che colla scorta di una buona e legittima induzione, e con certi raziocini, che in mercato non si vendono, tengono per certa la gravità de' corpi rispettivi di Marte, di Venere, e degli altri Pianeti."

78. SLI 2 (1751): 137.

79. SLI 3 (1752): 271. They referred the reader to Boscovich, *De centro Gravitatis Dissertatio* (Rome: 1751).

80. SLI 5 (1754): 113. These efforts are examined by Charles Coulston Gillispie, *Science and Polity in France at the End of the Old Regime* (Princeton: Princeton University Press, 1980), 112-13.

81. SLI 3 (1752): 257: "Perchè dunque da' sopraddetti autori con questo metodo non è stato sciolto?" The work in question was *Compendiaria Dissertatio de coloribus . . . pars prima* (Florence: 1749).

82. SLI 2 (1751): 156. Pemberton's work was translated as *Saggio della filosofia del signor cav. Isaaco Newton* (Venice: 1733, reprinted in 1745).

83. SLI 3 (1752): 212. The issue is covered in Newton, *Opticks*, II: iii: 8. Ferrone discusses the Italian side of the quarrel in *Scienza natura religione*, 250-6. The Transalpine debate is covered by Henry Guerlac, *Newton on the Continent* (Ithaca: Cornell University Press, 1981), 78-164.

84. SLI 6 (1754): 130: "I Newtoniani ormai non degnansi più di provare, che la luce sia un effluvio da' corpi lucidi mandato fuori; lo suppongono siccome indubitata cosa, e quindi passano a spiegare i curiosi e vari fenomeni della luce. Eppure di grandissimo incomodo è l'esplicare in questa sentenza la successiva, ma oltre ogni credere velocissima, propagazione della luce in spazi sì vasti, e lontani; al che sarebbe necessaria cosa o mettere nel corpo lucido forze maggiori di quelle che sogliono o possono da' corpi cacciare cotali particelle, o nelle trasmesse particelle fingere un affatto incredibile tenuità, per la quale ancora con leggerissimo impeto potessero con tutta celerità a tanto immensi spazi venir mandate." The work in question was Niccolò Arrighetti, *Lucis Theoria* (Siena: 1752).

85. Hooke's theories and their relation to Newton's are examined by Patri Jones Pugliese, *The Scientific Method and Mechanical Investigations of Robert Hooke*, 2 vols., Ph.D diss., Harvard University, 1982, 2: 592-617.
86. SLI 3 (1752): 263. The work in question was *Electricorum effectuum explicatio, quam ex principiis Newtonianis deduxit, novisque experimentis ornavit D. Andreas Bina Mediolanensis* (Padua: 1751).
87. SLI 3 (1752): 267.
88. SLI 5 (1754): 181. The work in question was Pietro Cornacchini, *Lettere fisicomediche* (Siena: 1751).
89. SLI 8 (1755): 70-72.
90. SLI 1 (1750): 57-58. The witch debate is covered in Venturi *Settecento riformatore*, vol. 2: *La chiesa e la repubblica dentro i loro limiti*, 355-410.
91. SLI 8 (1755): 73: "Il trattar degli Enti invisibili appartiene non meno al filosofo, che al teologo, mentre la pneumatica parte della metafisica è destinata a un simile trattato, e che quando in tal materia questi due sono tra di loro discordi, se il teologo produce dottrine dubbie, e il filosofo dottrine manifeste, quest'ultimo è in diritto di pretendere, che il primo debbasi seco lui accordare." I compare Ferrone, *Scienza natura religione*, 273.
92. SLI 8 (1755): 73. Benedict XIV's scientific interests are analyzed in Gross, *Rome in the Age of Enlightenment*, 239-41.
93. SLI 3 (1752): 221. "Sino a tanto che fiateremo, viva Dio, non lascerem mai di condannare gli errori che la Chiesa Romana riprova, e d'opporci a chiunque e' sia, e 'n qualunque modo il faccia, il quale cercasse di promuoverli e di ristabilirli." Ludovico Barbieri, *Nuovo sistema intorno l'anima delle bestie* (Vicenza, n.d.), is analyzed in SLI 3 (1752): 275-78. Genovesi explained his position in *Elementa metaphysicae* (Naples: 1743). The whole quarrel is finely examined by Maria Teresa Marcialis, "Meccanicismo e unità dell'essere nella cultura italiana Settecentesca," *Rivista critica di storia della filosofia* 37 (1982): 3-38.
94. SLI 1 (1750): 122.
95. SLI 1 (1750): 122: "Credendo ciascuno di aver diritto di filosofare su tali sperimenti, forma da sé nuovi sistemi, inventa nuove ipotesi, e involge in maggior oscurità la ricerca de' veri principi.
96. SLI 1 (1750): 122.
97. SLI 6 (1754): 128: "Furono fatte all'autore alcune obbiezioni. . . . Ma forse le nuove difficoltà che potrannoglisi fare gli serviranno perchè meglio si spieghi e frenando i trasporti del fervido suo ingegno disamini anche con maggior cura le materie, che restangli a trattare."
98. SLI 3 (1752): 233.
99. Gian Vincenzo Patuzzi, student of Concina, ["Eusebio Eraniste"] *Lettere teologico-morali in continuazione della difesa della storia del probabilismo e rigorismo* (Trent [Venice]: 1751); by Macchi and others: *Lettere di ragguaglio di Rambaldo Norimene al suo dilettissimo amico D. Luigi Bravier intorno ad alcune controversie letterarie suscitatesi in varie città dell'Italia* (Trent [Lugano] 1754).
100. *Novelle letterarie* 12 (1751): 291-97.
101. *Theologia christiana dogmatico-moralis* (Rome: Simone Occhi, 1751), lxi.
102. *Novelle letterarie* 11 (1750): 139, 567.
103. *Appendice al trattato dell'uso del mercurio sempre temerario in medicina in giustificazione di Lorenzo Gaetano Fabbri, lettore di medicina nel gran ospedale di Firenze* (Lucca, 1751), 228.
104. *Novelle letterarie* 13 (1752): 360.
105. *Supplemento* 1 (1753): 225 (magic), 251 (lightening); 2 (1754): 228 (Sanvitali). Sanvitali responded in *Annali letterarie d'Italia* 1 (1762): 90.

106. *Epistolario*, 3: 79, letter to Bottari dated 24 April 1753.

107. Scipione Maffei recorded the incident in *Epistolario* 2: 1369, to Benedetto Bonelli, 5 Aug 1753: "Il generale de' Gesuiti fu talmente uffiziato alcuni mesi fa da quello de' Dominicani, che proibì al p. Zaccaria di continuar la sua Storia; ma egli si difese e la proibizione svanì." Zaccaria complained about these restrictions in *Difesa della Storia letteraria d'Italia e de suo autore contro le Lettere teologico-morali di certo P. Eusebio Eraniste...* (Modena: 1755), 114. The freedom of fellow-Jesuits elsewhere is analyzed by Antonio Acerbi and Massimo Marcocci, eds., *Ricerche sulla Chiesa di Milano nel Settecento* (Milan: Vita e pensiero, 1988).

108. Rosa, "Pubblicazioni e tribolazioni," 35, quoting a letter of 26 July 1756.

109. Rosa, "Pubblicazioni e tribolazioni," 40, reported by Ricci in a letter dated 22 July 1758.

110. Rosa, "Pubblicazioni e tribolazioni," 38, in a letter dated 1 July 1758.

111. The exhaustive work on Jesuit conspiracy theories is Alexandre Brou, *Les Jésuites de la Legende*, 2 vols. (Paris: Retaux, 1906).

112. *Lettera d'un cavaliere amico fiorentino al reverendissimo padre Lorenzo Ricci, generale de' Gesuiti esortandolo ad una riforma universale del suo ordine* (Lugano [Venice]: 1762), quoted in Franco Venturi, *Settecento riformatore*, vol. 2: *La chiesa e la repubblica dentro i loro limiti*, 20, my source for this and the previous paragraph.

113. *Elogio del Cavalieri* (Milan: 1778), 37-38. Pascal explored similar themes in *Lettres Provinciales*, letter 18, which I read in the edition by Hugh Fraser Stewart (Manchester: Manchester University Press, 1920), 244; La Chalotais explored them in *Compte rendu des Constitutions des Jésuites* (Rennes: 1762), 177-81.

114. On this point, the interpretation of Franco Venturi, *La chiesa e la repubblica dentro i loro limiti*, agrees with what is still the most detailed discussion still available, in Ludwig von Pastor, *The History of the Popes*, 39 vols., trans. E. F. Peeler (London: Routledge and Kegan Paul, 1951), vol. 38, passim. Critiques of the positivist account of eighteenth-century science and religion are in Gianvittorio Signorotto, "La devozione settecentesca. Tradizione e mutamento," *L'editoria del Settecento e i Remondini*, ed. Mario Infelise and Paola Marini (Bassano: Ghedina e Tassotti: 1992), 183-95; and Vincenzo Ferrone, *I profeti dell'Illuminismo*, passim. Consider also John Hedley Brooke, *Science and Religion. Some Historical Perspectives* (Cambridge: Cambridge University Press, 1991), chaps. 1-5.

115. "Lo stato presente della letteratura italiana," *Nuovo giornale letterario d'Italia* 1 (1788): 60-64, beginning an article that extended over several numbers, transcribed in Berengo, ed., *Giornali veneziani*, 618-26.

Chapter 7:
Science and the Marketplace

Not all observers were equally enthusiastic about the agricultural academies that had begun to spring up in various parts of Italy after the mid-eighteenth century. "Endless chatter" was their main contribution, in the view of Carlo Antonio Pilati, writing in the 1760s. That these typical assemblies, newly created or else grafted on to the aging trunks of previously existing academies, were occasionally the sites for scientific discussions, made no difference. They were nothing but places where "a dozen good for nothings without a rood of land among them, teach the proprietors of estates . . . a few hundred idiotic rules about an art whose single most important precepts are industry and diligence."[1] So from his point of view, Andrea Memmo and Andrea Tron in Venice were as unworthy of praise as was Ubaldo Montelatici in Florence for their contributions to transforming a Renaissance institution into an instrument of the Enlightenment. For Pilati, such academies were of no use for solving the serious problems in education, landholding, religion, and government that threatened to keep the Italian states in the backwaters of Europe.

And there is no doubt that the academies failed to serve as incubators for the liberating ideas endorsed by Pilati and the more radical proponents of the Italian Enlightenment. However, there is also no doubt that the joining of natural science and economics, mathematics and animal husbandry, in a public setting, marked an important step in late eighteenth-century culture.[2] Pressing problems of alimentation and public health imposed. The new academies helped encourage the growing demand for more accurate and incisive applications of science to practical problems. As such, they contributed to the late great eighteenth-century resurgence in the age of Spallanzani, Galvani, and Volta and the joining of science to technology. Let us explore the roots of this late eighteenth-century trend.

It did not take a Pilati, or any other particularly acute thinker in the 1760s to realize that the various societies on the peninsula were struggling to cope with the economic and social consequences of biological disaster. If ever an opportunity presented itself for the application of science to society, that time was when famine struck throughout Italy, along with its inevitable accompaniment of disease. In Lombardy alone, the price of grain rose some 60 percent in a single year.[3] Bread riots were recorded in the kingdom of Naples and the Papal States. Even where the crisis was experienced with less intensity, as in Venice, its echoes were profoundly felt. And anyone who was not aware of it first hand needed only to look at wave after wave of publications describing the dismal state of the cities and the even more depressed conditions in the countryside.[4]

To be sure, little could be done about these conditions in any emergency unless the basic systems for distributing foodstuffs could be corrected. Economic policies governing relations between cities and their hinterland were in some ways modeled on the mercantilist relations between one state and another.[5] Strict control and planning were regarded as the only bulwarks against chaos and confusion. Yet, commandeering supplies from the countryside did not always suffice to guarantee that city warehouses would remain full. That grain could be forcibly purchased by governments at below-market prices was hardly an encouragement to produce more efficiently. The same went for the government subsidies often paid to bakers when the price of grain rose too high. Moreover, tremendous powers in the hands of the governing elite who controlled the distribution system were an incentive to bribery and corruption.[6] And no amount of protests by government officers and their clients sufficed to remove suspicions that the main causes of scarcity were speculation and hoarding by the gentry with whom they were in cahoots.

Economic science emerged in part in reaction to the debacle of mercantilism in the grain trade.[7] Gradually, during the course of the 1760s, there began to take root the conviction that the most effective solution to the practical problems and misunderstandings in the area of food distribution might be to scrap the entire provisioning system and introduce a free market. Ideas that were discussed in a general way by Ferdinando Galiani were applied to the specific case of Naples by Antonio Genovesi and Emilio Coppa.[8] Projects proposed by Pompeo Neri, Antonio Serristori, and Giambattista Uguccioni in Tuscany were discussed by the physiocrats in France. Finally in Tuscany freedom of the grain trade was decreed in 1767, and the same was decreed two years later in Venice.[9]

Expectations for change along these lines were of course tempered by the skepticism of the proponents themselves concerning the capacity of populations to adjust. They also ran up against the stiff opposition of entrenched groups—in Naples, protected to some degree by the otherwise enlightened Bernardo Tanucci and in Rome by the much less enlightened Consoli d'agricoltura.[10] Of course, a free market in grain was not much use to correct the devastating effect of the traditional distribution of land, or the traditional contracts of tenancy. And the general questions taken up by Pilati were applied to Naples by Filippo Villani and to Rome by Claudio Todeschi. In Florence, to Leonardo Ximenes' complaints about feudal laws and fiscal impositions were added those of Giovanni Lapi concerning the sharecropping system.[11] Suggestions for land reform met with less success in Tuscany than in Venice. In Venice, the issue of landholding was connected to the issue of Church property; and this was addressed in a series of remarkable projects championed by Andrea Tron and Andrea Querini. But even here, progress was extremely slow and virtually stopped after 1774.[12]

The application of science to agriculture seemed to promise less socially disturbing solutions and perhaps some shorter-term gains than more incisive reforms. There were, after all, other ways to get higher yields and better crops than by dismantling the old regime. Ubaldo Montelatici, in collaboration with the Georgofili and Pietro Leopoldo of Tuscany, suggested the reclamation of the Maremma of Siena.[13] Others recommended better medical and scientific supervision over foodstuffs brought in from possibly infected places. Saverio Manetti called for a census of grain types, and for a new science of alimentation using historical and ethnographic evidence to discover how humans might nourish themselves cheaply without depending entirely upon wheat. Giovanni Targioni Tozzetti was so impressed by the convergence of intellectual resources that he declared himself satisfied that the Tuscan countryside had been placed on "the most solid foundations of experience" and guided by "chemical and physical discoveries."[14] He himself advocated further crop variation and hybridizing.

All of these initiatives suggested a new level of interdisciplinary collaboration. Not that the debate about the application of science to public welfare had been entirely absent from the Italian intellectual scene before this time. Already in 1749, Ludovico Antonio Muratori, in his *Della pubblica felicità oggetto de' buoni principi*, insisted on medicine and agriculture as the foundations of public welfare. "The study of nature," he noted, "intent as it is on discovering the arcana of God's creation, may greatly help medicine, agriculture, economics, navigation, human commerce and so many other needs and conveniences of our lives."[15] And he praised the cities of Paris, London, Berlin, Petersburg, and even Bologna (referring to the Bolognese Institute) for having established academies for this very purpose. He only wished that more natural philosophers had turned their attention to agriculture, by conducting experiments on the cultivation of the earth, on certain plants, and, especially, on pesticides.

Giovanni Lami continued Muratori's campaign, in his *Novelle letterarie* literary journal. To Muratori's proposals, he added his own insistence on the connection between science and the arts of commerce. Manual labor, he proclaimed, ought not to be scorned by those who concern themselves with the realm of ideas. On the contrary, let philosophers learn some of the techniques of the trades, even the operations of some machines.[16] The precision tasks of clock making, for instance, might be particularly appropriate for those who had to perform delicate scientific experiments. Turning tools on a lathe might give the researcher the insight needed for designing his own laboratory instruments. Finally, familiarity with the trades might inspire researchers to discover new applications of their own knowledge. As an example of a researcher who had successfully united theory and practice, philosophy and usefulness, he cited René-Antoine Ferchault de Réaumur, who in the course of his work at the Paris Academy of Sciences came up with a new and cheaper method for making porcelain. At least for a time, he praised the editors of the French *Encyclopédie,* a veritable *dictionnaire raisonné* of the arts and crafts.[17]

Antonio Genovesi, the holder of the chair of political economy at the University of Naples, tied economics to science and technology in many of his productions. He debated John Locke's assertion that physics was a speculative art. Physics badly understood, he agreed, was never useful to humanity. But good and true physics always was. "Aerometry, hydrostatics, all of statics and mechanics, are of very great importance to animal life, and for the arts and for commerce."[18] To prove his point, he provided an edition of Petrus van Musschenbroek's famous physics textbook, the *Elementa physicae*, as well as an edition of Cosimo Trinci's *L'agricoltore sperimentato*.[19] The misery of the Kingdom of Naples, he suggested, was partly due to the absence of collaboration between specialists in different disciplines. Instead, let philosophers avoid "sterile contemplations" and, ignoring the social stigma attached to the "mechanical arts," let them turn their attention to "agriculture, the theory of commerce, natural history, mechanics and similar very useful sciences."[20] Like Muratori, he viewed academies as the places where the best-coordinated effort was likely to go on. In general, where the sciences "are true and lucid, and diffused among the public, men live better by them and learn to do much with little exertion."

In the 1750s, a major breakthrough in the application of advanced research to problems of public welfare was the campaign for smallpox inoculation. Inspired by Charles Marie de la Condamine's widely circulated tract, a significant public relations effort sought to break down popular resistance.[21] Careful instruction drew attention to the methods of modern medical research, the use of statistics for determining the effectiveness of a chosen therapy, and the standards of professional behavior. Finally suspicions about the danger to lives and souls inherent in the use of material from diseased tissues began to subside. Even the Church hierarchy was convinced, including Pope Benedict XIV; and a number of political authorities took the inoculation themselves as examples for others. The successful inoculation campaign in Florence was described by Targioni Tozzetti in 1757, and that in Siena was described by the authors of the *Atti dei fisiocritici* in 1760.[22]

The crisis of the 1760s demonstrated that the initiatives to date in the realm of the application of science to society were not nearly enough. Epidemics still spread with frightening rapidity through the run-down shacks in which many people lived at the lower end of the economic spectrum. And even the few who managed to get their smallpox vaccination were exposed to pleurisy and the other myriad diseases associated with miserable conditions.[23] Hospitals were understaffed and overcrowded. However, among some intellectuals, the combination of these complex health issues with basic problems of nutrition did not lead to total despair. A growing faith in the infinite perfectibility of mankind, reinforced by small but steady steps in the accumulation of knowledge, boosted hopes for change. These hopes in turn occasioned the most remarkable episodes of intellectual and cultural organizing in Italy since the Renaissance.

The Venetian Republic was a good example of the trend. On the model of Montelatici's Accademia dei Georgofili in Florence, the local literary academy of Udine transformed itself into an agricultural assembly after 1762; and that of the Risorti in Capodistria followed suit. And while discussions about liberalizing commerce in grain echoed throughout Italy, the Venetian government undertook a radical plan for focusing expert attention on problems of alimentation and environment. Not satisfied with simply admitting the inadequacy of current methods of cultivation in the face of rural demographic growth, the government appointed the professor of botany at the University of Padua, Pietro Arduino, to a new chair of agronomy. It paid close attention to his report that "the disorders currently abounding in the old commonly used practices are almost innumerable."[24] Next it appointed Pietro Arduino's brother Giovanni, a well-known geologist, as superintendent of agriculture and advisor to a new government magistracy called the Deputati all'Agricoltura.[25] And in all the cities of the Terraferma, government-funded agricultural academies were either set up (as in Cologna Veneta) or grafted on to the trunks of existing academies (as in Verona).

To be sure, most of the academicians dedicated themselves mainly to studying methods for improving profits on their own land holdings rather than to those for improving the lives of the inhabitants of the countryside.[26] Recurring themes in academic discussions were therefore the advantages of the three-field system, the division of communal property, and the protection of arable land. And much of their rhetoric about studying "discoveries, experiments, machines, and other similar things, whatever may contribute to the great field of agriculture," was just that: rhetoric.[27] However, they also experimented with the introduction of new crops such as hemp and madder for industry; and they participated in government competitions to discover new methods of raising livestock, offering their own prizes for more efficient methods of preventing mulberry and olive plant diseases. They even saw some of their suggestions turned into law when the Senate promulgated legislation to prevent deforestation or to control grazing.

For all of these reasons, the academies deserved the praise of Antonio Zanon, who in response to Pilati's criticism, asserted that they "were among the most useful institutions" he knew. After all, he added, their subject matter involved the most pressing concerns in all of Italy. And "what matter of any importance was ever concluded without the collective effort of people assembled together?" No wonder their discussions had already produced "innumerable advantages," not only to the science of economics, but also to manufacturing and commerce, and all those aspects of human existence that made states populous and flourishing.[28] Cesare Beccaria went even further. To him, "agricultural academies are among the most useful things known to the human race." Did they not "submit" agriculture "to physics, to mechanics, to chemistry?"[29] From these sciences philosopher-farmers might learn their trade more thoroughly and then set an example for the peasants. Every important human endeavor, he believed, must be pursued in an orderly fashion if it is to succeed. Order, moreover, is a product of the application of physical as well as moral laws. These laws are not always self-evident; and their discovery requires study and effort. Once discovered they may be communicated to those who need them most. Academies were particularly suited to this task.

In spite of the critics' warnings, the Venetian academies met with surprising success. They inspired Luigi Riccomanni to propose similar academies for the Papal States, with explicit reference to Venice.[30] They inspired the Venetian Senate itself to extend their purview, in 1779, to all of commerce and industry. After all, they claimed to pursue "the best means for keeping alive the passion for the good of mankind in the hearts of young people who will one day be powerful and rich."[31] If they engaged their members in competitions for designing a better type of lathe or discussed mew methods for constructing violins and other musical instruments, as the Academy of Padua did in 1786, such activities were fully in agreement with their dedication to public service.[32] No wonder they were finally transformed into the academies of "science, letters and arts" that survived the Napoleonic period to carry on a distinguished career of intellectual development throughout the nineteenth century.

But from the start, the academies offered, and, more importantly, the academies encouraged, something more than solutions for improving agriculture. They contributed a considerable impetus to discussions about joining intellectual interests to practical ones, theory to practice, and science to technology. Indeed, by the 1770s, an authority like the papal governor of Ferrara, Claudio Todeschi, called upon governments to "prefer the more useful sciences to the merely speculative ones."[33] There was already considerable precedent for such sentiments. Antonio Genovesi was not the first to suggest, "everything in the sciences that is not useful for mankind is a waste of time." He only expressed it most forcefully. "If philosophy has helped us in anything," he went on, "it is precisely this: in having disabused us of so many useless applications of our forefathers." In his view, natural history was the particular field which, pursued at the expense of other more abstract studies, was most likely to "make Italy less unhappy."[34] As a solution to humanity's problems, Voltaire, at least in Antonio Zanon's recollection, joined to this the study of physics.[35]

In fact, the academies were as much fomenters as symptoms of a major change in what might be called the public sphere of science in the latter half of the century.[36] And the publications that purported to cover their activities, such as the *Giornale d'Italia* run by Francesco Griselini or the *Magazzino toscano*, run by Saverio Manetti and collaborators, helped turn the campaign for more useful applications of science and technology into a commonplace. The latter, in fact, apart from its insistence on an agricultural practice "based on the most solid foundations of natural science," extended its purview to many topics connected with economic improvement and public welfare. A series of articles on smallpox advocated the methods of Thomas Dimsdale. For rabies, a mercury-based medicine was reported as having been invented by a Jesuit chemist at Pondicherry. And the application of chemistry to enameling processes was referred to the recent discussions about the phlogiston theory. Invitations were extended to a broad array of possible contributors on technical subjects. And, the editors added, "Literary figures who favor this work with their dissertations or their new discoveries in every branch of science will not be disappointed."[37]

The veritable flood of general audience journals and magazines that came off Italian presses in the 1760s and 1770s, from the *Caffè* in Milan to the *Giornale de' letterati* of Pisa, covered the latest developments in Enlightenment science from the standpoint of its applications to society. And when they discussed agriculture, they placed their opinions in the context of all the other areas of natural knowledge that were in any way connected with it. Noted Giovanni Francesco Scottoni, in his Venice-based *Memorie utili*, "Agriculture is a branch of natural philosophy, and not only one of the most useful, but one of the vastest."[38] Yet in his journal, "all the most significant discoveries that might be useful to the arts, the trades and manufactures will be noted."

There is no way of measuring precisely how much the widening public sphere of science contributed to changing the research programs of practitioners in the latter half of the eighteenth century. A number of specific areas of study had developed from discussions in the realms of natural knowledge over the previous half-century. Some of these lent themselves particularly well to the kinds of inquiries that were being made by those who promoted science's public role. Problems concerning the generation of human life owed as much to the tradition of Malpighi and Vallisneri as they did to the new demand for a healthier population. Problems concerning the quality of air, earth, and water owed as much to the tradition of Agostino Scilla, Guglielmini, and Poleni as they did to the new demand for a safer and more productive environment. Research on electricity owed as much to the pioneering work of Eusebio Sguario as it did to the search for the secrets of human life. However, the itinerary from these disparate elements to a new science of man by way of the most exciting discoveries of the age can be roughly traced in the careers of a few important figures. We will consider four: Giovanni Targioni Tozzetti, Alberto Fortis, Lazzaro Spallanzani, and Luigi Galvani.[39]

In the case of Targioni Tozzetti, the imprint of the public sphere upon scientific productivity appears particularly clear. Trained in medicine at the University of Pisa, he was introduced to natural history and especially to botany by Pier Antonio Micheli.[40] While continuing to practice medicine in Florence, he took over instruction at the botanical garden upon Micheli's death, and soon became grand ducal librarian. This official post placed Targioni Tozzetti in something of an advisory position within the grand ducal system. We have seen some of the consequences of this in his work on inoculation for smallpox. His early work on agriculture reflected the same professional commitment.

Throughout this period Targioni Tozzetti accepted official missions into the Tuscan countryside to identify mineral deposits and geographical characteristics that could be exploited for public revenue. And partly as a result of this activity, from 1751 to 1754, he published the four volumes of his reports on voyages through Tuscany, which he was to revise and increase in later years. It was a monumental work that drew his talents to the attention of an international audience. In the 1750s he planned an exhaustive naturalistic, geographical, and hydrological description of Tuscany. The preliminary work he did on the project went a good way toward defining the standards in this genre.

In the 1760s, with famine and epidemic raging in Tuscany, Targioni Tozzetti's productivity shifted largely into the fields of applied science. In one work, he studied the relation between hygiene, climate, and contagion.[41] In another, he examined the causes of crop failure, attributing the Tuscan case to the spread of wheat rust. He thereupon analyzed the life cycle of the parasite responsible for this disease.[42] In yet another production, printed and distributed at his own expense, he considered the nutritional value of alternative forms of alimentation that could be used in times of famine, including flours made of ground beans, garbanzos, and peas, and even the pods of beech trees. Using his knowledge of human geography, he recorded occasions when imported crops had been adopted around the world with beneficial effects, advocating the cultivation of the potato, as well as diverse varieties of wheat deriving from different areas and climates.[43] In his mind, science and application went hand in hand.

Targioni Tozzetti's lessons on social geography were carefully heeded by Alberto Fortis. Just in the period when the Venetian Republic's agricultural academies were gathering momentum, and Targioni Tozzetti's more utilitarian works were coming off the presses, Fortis completed a research trip to the Venetian colonies in Dalmatia. There, he was as stunned by the variety and richness of the chorography as he was scandalized by the primitive economic and cultural circumstances of the residents. Ignorant and superstitious, they had no concept of extending their fishing businesses into the Venetian markets; and their agricultural implements had remained unchanged since antiquity. Nor did the Venetian government seem to see the advantages of viewing them as anything but a source of tribute.[44] Fortis' suggestions implied a pioneering ecological approach, calling for reforestation, experimentation with new crops, and advocacy of small proprietorship to focus farmers' efforts on land improvement.[45]

Like Targioni Tozzetti, Fortis combined his scientific work with an intense public commitment, attuned to the practical questions that were being raised in the new information media. Trained in natural science at the University of Padua under Antonio Vallisneri, jr., the son of the great naturalist, he turned his back completely on his maestro's outdated attempts to shore up the humanist or neoclassical underpinnings of the Paduan tradition of natural science represented by Vallisneri senior.[46] Instead, he came under the influence of Giovanni Arduino, the Venetian government's specialist on geography. His scientific activity took place in the midst of writing about political and social subjects of current interest for the *Giornale d'Italia* and another general audience periodical published in Venice, the *Magazzino italiano*.

Drawn to the kingdom of Naples at first by the prospect of grounding new hypotheses about the origins of the earth in the abundant evidence of volcanic activity, Fortis soon tuned his attention instead to the problems of the Bourbon monarchy there. After he came into contact with the circle of the Enlightenment jurist Gaetano Filangieri, what began as a geological survey eventually became a search for new resources that might bolster the faltering economy of the south of Italy. The discovery of significant saltpeter deposits in Puglia appeared to promise not only employment for families but also financial benefits that could solve some of the government's fiscal needs and encourage reductions in taxes. While engaging in a scientific debate concerning the value of his discovery, Fortis undertook a courageous, though ultimately unsuccessful, crusade to gain government backing for what would have been one of the most significant public mining enterprises in southern history.[47]

Like many in his time, Fortis saw electricity as a promising new tool in the shop of the enlightened philosopher, for solving some of the urgent problems of public health and human welfare. There was considerable precedent for this view. Electrical research in Italy had become a regular part of instruction in experimental natural philosophy already in the first third of the century; and friction machines, electroscopes, and Leyden jars were still being imported from northern European manufacturers and utilized according to the instructions found in textbooks by Willem Jacob 'sGravesande and Petrus van Musschenbroek. While researchers such as Eusebio Sguario in Venice and, later, Giambattista Beccaria in Turin, had concentrated on the nature of the electric fluid and the physical causes of its effects, much attention had been devoted to electricity as a branch of medicine.[48]

A convenient summary of the medical potential of electricity had in fact been published in the Venice of Fortis' time. The author was Giovanni Francesco Pivati, a Venetian attorney whose views owed less to the physics of Sguario and Beccaria than to the electrical cures for paralysis attempted by Jean Antoine Nollet, the French master electrician. Surely, Pivati noted, such work offered the best hope of bringing medical science out of the obscurity in which it had languished since professional prejudices had separated theorists from practitioners at the origins of the modern age. After recounting what had been tried by others in the way of cures, he cautiously included his own. Although no definitive results had yet come from having patients hold electrified cylinders containing medicines, Pivati was certain that such experiments would reveal entirely new ways to deliver curative substances without disturbing the veins.[49] Fortis agreed.

The mysterious aspects of electrical medicine, which seemed to defy not only the familiar physics of Isaac Newton but also the anatomical theories of the school of Morgagni, in the midst of endless frustrations and dead ends, drew Fortis toward some of the more speculative work that was being done in the field at this time. Rather than to the theories of the physicists, of Benjamin Franklin or Albrecht von Haller or even Nollet, Fortis gravitated toward those of the controversial physician Pierre Thouvenel, whom he met in Venice. The similarity of some of Thouvenel's views to those of the equally controversial Franz Mesmer, including the use of hypnotism in therapy, only added to their appeal. To Fortis, the notion of a single cosmic fluid, coursing through all bodies, organic and inorganic, animating them, seemed to carry enormous promise, not only in medicine but also in any other areas, such as dowsing, where animal electricity might be involved.[50] The unfulfillment of these hopes fueled the ardor of his faith in a future age of progress.

That Lazzaro Spallanzani should, at least at first, have shared Fortis' enthusiasm for the theories of Thouvenel, says much about the seductiveness of even the most extravagant views, when so much was at stake in the medical and social worlds.[51] Spallanzani was drawn to the problem of the nature of electricity less by a quest for the moving force in the universe than by its bearing on questions concerning the movement of the muscles and the actions of the nerves. So much, in his view, depended upon going beyond the older mechanistic explanations of such phenomena, that he was willing to entertain almost any possibility, provided that it could be backed up by empirical evidence.

In his earlier work, Spallanzani followed Haller in attributing muscular motion to an irritability in the tissues, stimulated by the action of the nervous fluid. Eventually he abandoned this view for the theory that motion came from an electrical force inherent in bodies. As long as Thouvenel and his followers appeared to rely on legitimate experiments and not on prearranged demonstrations, their ideas might offer useful working hypotheses concerning the nature of life itself.[52] When this was no longer the case, and some professional malpractice seemed to be afoot, Spallanzani found more congenial support in Galvani for his ideas on the subject.

Fitting Spallanzani's highly varied activities into a coherent pattern is no easy task. He was a biologist as well as a physiologist, a natural historian as well as a geologist. The absence of any specific programmatic statements in any of his work need not be taken as a sign of his disinterest in the wider implications of what he was doing.[53] Nor should his ordination into the priesthood be construed in itself as proof of a specifically religious view of nature. However, the picture we have been tracing here of the late eighteenth-century cultural environment provides some highly suggestive clues. If we take Spallanzani as the product of the same world in which Fortis and Firmian, Pietro Verri and Bernardo Tanucci, shared passions and interests, a single theme seems to stand out: the search for a useful scientific understanding of the vital processes in which humankind was involved.

By examining the nature of life—or, as he put it, of the vital heat in living beings—Spallanzani reached to the most basic level of a science of man. One project investigated the ingestion of nutrients. Ever in search of the uniformities that situated humans within the vast scheme of nature, Spallanzani set out to correct the defective observations of others by his own innovative, exhaustive and sometimes painful experiments. And having discovered the life-giving fermentation process in dozens of species, he explored it heroically in his own body by swallowing and disgorging packets of food to observe the relative roles of gastric fluids and peristalsis.[54]

The spontaneous generation controversy inspired Spallanzani in a series of productions with wide implications for the beneficial control of life forces, mostly unrealizable at the time.[55] The beliefs that had drawn him to Thouvenel's theories concerning a universal source of vitality did not extend to others' theories about the universality of life. For Spallanzani, the notion that inert matter could somehow be vitalized conflicted with what he knew about the reproduction of species. He attacked John Turberville Needham's supposed demonstrations of the spontaneous generation of infusorians by far more exhaustively repeated experiments showing the conditions in which these organisms, which would be the objects of sterilization practices in the following century, grew and survived. He applied equal diligence to the role of insemination in amphibians and mammals, performing artificial insemination experiments that would have wide implications in animal husbandry.

More immediately applicable to practical concerns was Spallanzani's research on the quality of the air breathed in various parts of Italy. The question itself had emerged amid increasing concern about illness due to the atmosphere in mines and swamps.[56] The discovery of the basic composition of air had led to the invention of eudiometers to measure the life-giving portion of it—i.e., "vital air" or oxygen. Spallanzani did not set out to dampen hopes for fresher air indoors and outdoors, but to set the therapeutic issues on a firmer chemical basis. He distinguished the natural life-giving component of the air from the occasional pathology-inducing pollution that might accompany it. Contrary to many popular self-proclaimed respiration specialists, he was able to show that the air around a place like Modena, a city notorious for its swamps and fevers, was as full of "vital air" as was that of any city of Italy. Variations in "vital air" or oxygen content could instead be found very clearly between different heights above sea level. And in his sample-gathering procedures and laboratory record keeping he gave examples of scientific rigor to this newly developing field of research.[57]

Finally, Spallanzani studied the ways in which the physical environment affected human settlements. As well as a treasury of observations, he amassed a peerless collection of natural specimens, with which to enrich the public museum of Pavia.[58] Like Fortis and Targioni Tozzetti, he provided thorough descriptions of his natural history expeditions, combining sociological observations with zoology, botany, and mineralogy. If plants and animals were best observed amid their surroundings, so too were men.[59] And while using fossil evidence to disprove notions about the role of the biblical Flood, and testing volcanic processes by experiments in the laboratory, he turned a naturalist's eye to problems of human as well as natural disaster. In Messina he could not conceive of how, four years after the last eruption of Etna, "a good part of the private and public buildings are still in the same pitiable state in which they were left during that unfortunate period; so that many people are still forced to take refuge in their ruined houses, in filthy hovels and in flimsy sheds. And nearly all are still oppressed and weighed down by fear, and, one might say, by degradation."[60]

For keeping his interests focused on the practical side of science, Spallanzani, like Galvani some five years later, benefited from the cultural environment of the city of Bologna. In Spallanzani's case, the guidance of his cousin Laura Bassi played a crucial role, while he completed his degree before proceeding to Reggio Emilia and finally settling at the University of Pavia. It was enough that she was one of the more inspiring figures on the Italian intellectual scene at this time, a correspondent of Morgagni as well as of Voltaire. Spallanzani no doubt benefited from her well-attended public lectures in philosophy at the university. More importantly, she kept a famous home academy, where the practical aspects of experimental philosophy were discussed and demonstrated by the use of a considerable repertoire of electrical instruments and paraphernalia.[61] To her, Spallanzani dedicated his first major scientific publication, concerning the effect of water on rock formations.[62]

Again, for Spallanzani as well as for Galvani, the Bolognese Institute functioned as a center of advanced interdisciplinary research and instruction. From its very foundation in the early eighteenth century, it had been organized to facilitate interdisciplinary borrowing. Inside Palazzo Poggi, its official headquarters, the laboratories of anatomy, chemistry, physics, and so forth were placed in close proximity precisely in order to permit continuous interchange.[63] The subjects examined there ranged from animal respiration to chronometry; from the composition of water to atmospheric electricity; from fossilized wood to the medical uses of camphor; and early volumes of the Institute's *Commentarii* covered natural history expeditions by top members.[64] As the one Italian academy where rigorous scientific standards combined with a serious public role, the Institute could not be blamed for contradiction if its published proceedings were in Latin rather than in the more accessible vernacular of the *Giornale d'Italia* or the *Magazzino toscano*. After all, unlike the agricultural academies which these latter publications represented, the Institute organized a vernacular lecture series, where city elites as well as students could be expected to attend.

Chemistry and obstetrics were just two of the many new fields that were recognized at the Institute as concessions to the new useful science that Spallanzani and Galvani never tired of advocating. Long after Spallanzani had left Bologna, and even longer after Galvani had been appointed to the chair of anatomy, Galvani himself took over in obstetrics. In accepting the chair, was he truly "inspired by the desire to serve the public weal as much as possible through his study and endeavors," as his eulogist later wrote?[65] More precise information has not survived; nor can we be certain that he indeed paid as much attention to ministering to the poor and indigent as he is reported to have done, following his own counsel (to his nephew) to "make yourself dearer to your countrymen by your virtue, integrity, impartiality and toil."[66] One thing is certain. During the course of his career as the chief physician in Bologna, he had no shortage of opportunities for putting his services to work for public benefit—from advice on what to do in time of contagion, to counsels concerning a cow disease that spread through the communities of Vimignano and Savignano and elsewhere in Bolognese territory in the year 1775.[67]

For turning electricity from a kind of scientific parlor game into a promising new branch of public health, Bologna turned out to be the ideal place. The European debates arrived virtually at Galvani's front door; and he was able to acquire a complete collection of the most recent books on the subject.[68] Moreover, Giuseppe Veratti, a professor at the university and member of the Institute, was a particular proponent of the physiological uses of electricity. His technique, far simpler than Pivati's, was to apply electrodes at or near the affected area of the patient's body, in repeated sessions. He reported considerable success in curing arthritis, after delivering reportedly quite painful shocks to various limbs. To patients complaining of deafness, he administered shocks to the ear. Likewise, with patients suffering from migraine headaches, he found that shocks to the head were at least somewhat less uncomfortable than the malady itself. Some experiments he apparently carried out in his home along with his wife, Laura Bassi, for an audience of students and acquaintances. But he was never able to explain the reasons for his results, preferring instead the "bare and simple observations."[69] Fortunately for Veratti, his prestige was only increased when his claims were challenged by Jean Antoine Nollet, then in the midst of a campaign to unmask scientific fraud.[70]

When Galvani took up this research, debates about electricity seemed to have reached something of an impasse. Giambattista Beccaria's fundamental *Dell'elettricismo artificiale e naturale* of 1753 explored in depth the influence of electricity on the phenomena of life, including experiments tracing the effects of electricity on plant growth and muscle tension, with suggestions for curing paralysis, blindness, and muscle pains. In Beccaria's mind, all these experiments suggested that the nervous fluid was none other than an electrical fluid. In subsequent decades, the electrical theory of muscle contraction was widely proposed as an alternative to Haller's notion of irritability. Yet there was still no precise knowledge either of the nervous fluid or of the electrical one, and Leopoldo Marcantonio Caldani and Felice Fontana both found the electrical theory an easy target. As late as 1781, Fontana admitted, "O how many uncertainties there are to be resolved by our successors!"[71]

From 1780, Galvani began his own experiments in earnest. Already at this stage, he was convinced that animal electricity existed. On it, he said, "seem to depend movements, sensation, blood circulation and life itself."[72] In his search for the location of the electric fluid in the body, he undertook to respond to Haller's conjecture that it was not the same as the nervous fluid. He accordingly repeated Haller's experiments with tourniquets on animals to determine whether electrical impulses flowed when fluids were cut off. His results suggested that there was indeed a single fluid, and that it was electrical. Next he responded to Felice Fontana's suggestion that there was no way to explain how the nervous electric fluid circulated through the body. Galvani accordingly performed careful experiments on the interior of the spinal cord of frogs to show that in some way electrical impulses managed to travel through the nerve canals to the muscles in spite of the insulation.

By the time Galvani's research was in full swing, the Institute was enjoying a kind of mini-Enlightenment under the leadership of Sebastiano Canterzani. A mathematician and physicist, Canterzani included such figures as Condorcet and Pietro Verri among his circle of literary acquaintances. The latter sent him volumes of his *Meditations on Political Economy* and *Discourse on the Sources of Pain and Pleasure*, works that applied Enlightenment ideas to problems of public welfare. Together with other admirers of the French *Encyclopédie*, Canterzani formed a "Società Enciclopedica di Bologna" under the direction of Giovanni Ristori, who had fled to Bologna from Modena for reasons of censorship. The "Società" soon began importing foreign books on Enlightenment subjects as well as publishing works by members.[73]

Whether Galvani would have had much sympathy with the more audacious programs of the Bolognese Enlightenment is difficult to assess. Canterzani and Ristori were involved in one of the major episodes in Enlightenment publishing in Italy: the *Memorie enciclopediche*. A spin-off publication, the *Storia dell'anno*, delved into the questions that were troubling the Italian states due to "the prejudices of our century." To correct these, it gave cautious support to the politics of state centralism and anticlericalism characteristic of the Habsburg emperor, Joseph II, as well as to the latest stirrings of republicanism; it decried fiscal tyranny, called for better poor relief, revealed the social roots of crime, and recounted the story of the American Revolution according to Raynal.

The *Memorie* themselves proposed to bring the chief ideas of Enlightenment science, politics, law, and economics to a broad audience. "We wish to make ourselves understood by every sort of person," the authors insisted. "We do not write especially for the physician, the theologian or the mathematician. We wish to give useful pleasure to the gentleman and the lady, and every category of cultivated person." On page after page they referred to the major thinkers in the movement, in Italy and elsewhere, from Pietro Verri to Gaetano Filangieri, from John Locke to Joseph Hume, from Montesquieu to Rousseau. Voltaire, they noted, "joined all the talents of the others in himself." [74]

Canterzani and Ristori insisted far more than any of their contemporaries upon the uselessness of pure erudition—even in science. "Natural history is very useful," they admitted, "but what is the point of printing so many books to inform the public that a few shells were found in a certain four-*palmi* area of land . . . or that horse and donkey bones, or even those of an elephant, were found in some valley."[75] A good part of regular scientific investigation had been reduced to the same sorts of trivial pursuits that occupied the historical antiquarians. Some future writer of comedies, the authors of the *Memorie* assured their readers, would no doubt hold many self-important scholars of the age up to the most hilarious ridicule. Knowledge of any kind was only as good as its applicability to the real needs of mankind, as determined by the Enlightenment philosophers.

Broadly based social conscience periodicals such as the *Memorie enciclopediche* and the *Storia dell'anno* helped establish the terms on which the last crisis of the Old Regime in Italy, from 1789 to 1796, would be experienced by Bolognese citizens. This is not the place to consider the role of the Enlightenment in the social and political transformations at the end of the century. Nor is it necessary here to chronicle the hopes raised by the new republican government in France or the perplexities raised by the Jacobin policies there from 1793. The cause of reform in Italy received some of its worst setbacks in the six years preceding Napoleon's invasion, with the death in 1790 of Joseph II, ruler of Lombardy, and the departure of Pietro Leopoldo from Tuscany. The liberation of Bologna from papal control and its incorporation into the new republic of Cispadania, later in the Cisalpine republic, at first appeared to signal a new departure. In fact, the new Napoleonic world offered nearly as many opportunities for the play of self-interest and mischief as for social and political renewal.[76]

Galvani's politics, to be sure, were by no means revolutionary. Indeed, he would be among the university professors who refused to sign the loyalty oath to the new government in 1796. However, he joined his Enlightenment contemporaries in hoping for a bright new future guided by science—especially, a science of electrical medicine capable of prolonging life, diminishing disease and alleviating pain. Like them, he claimed to have been amazed by a "new" and "marvelous" phenomenon with enormous possibilities. And he awaited, in the words of his nephew Giovanni Aldini, "the arrival of that adventurous epoch when animal electricity would be drawn out of theoretical speculations and usefully employed for the relief of mankind."[77]

Galvani's impressions about the "new phenomenon" turned out, of course, to be misleading. He had already shown that muscles could be stimulated by the discharge of electricity from some external source. A crucial experiment now seemed to suggest that electricity stored in one part of a body could be drawn to another by a metal arc applied to that part and to a nerve, causing contraction.

He now suggested that the muscles were organic capacitors similar to Leyden jars, which maintained their charge until attached to a conductor. However, the new form of electricity, he thought, was of a different kind from that found in the atmosphere or created by machines. Surely this animal-based electricity was a peculiar animating fluid; it was, he believed, the essence of life. And in his view, it ought to be included among the other great motors of the universe: the calorific, the attractive, and the magnetic.

The possible uses to which his discoveries might be put engaged Galvani in the exercise of his lively imagination. Surely diseases like apoplexy and paralysis might now be understood, if not immediately eliminated. Animals had already been shown to experience violent convulsions followed by paralysis, when given electric shocks to the brain. Excess animal electricity in humans, discharging itself on the cerebrum, would have the same damaging effect. In such cases, the addition of further electricity could obviously be dangerous, so electrical cures were not indicated. Nor were they necessarily helpful in cases of paralysis. While proclaiming the explanatory value of his work, Galvani introduced a note of caution into research on new cures.

Galvani showed less reserve in his study of epilepsy. How could he ignore the analogy between the effects of electrical shock on individuals and the warning signs of a seizure? Unusual internal heat, increased secretions and excretions, sweat, and quickening of the pulse seemed to occur in both cases. Surely, again, epilepsy originated in a sudden discharge of electrical fluid. In diseases like rheumatism, on the other hand, perhaps the accumulation of stagnant electrical fluid in the humors might be the culprit. The administration of negative electricity would therefore be useful here, to remove the accumulation of the positive. Atmospheric electricity was a possibility, by grounding the patient and attaching a suitable conductor to the afflicted part. After all, some paralytics had reported sudden cures after lightening struck nearby. Which of these "facts" might be useful immediately, and which would be only "for future use" he left to his readers to decide.[78]

The impact of Galvani's discoveries on the scientific community was as powerful in Italy as it was in the rest of Europe. At the time, Gioacchino Carradori saluted the introduction of an entirely "new province" into the field of physics. As he later recalled, "at first all, or almost all physicists embraced Galvani's doctrine. All fought, as it were, under his standard, adopting and supporting this animal electricity by explanations and experiments."[79] Canterzani proclaimed that Galvani's discovery was not only new, but signaled the beginning of a "new epoch," while salons and academies filled with discussions of the issue. Learned journals and magazines set up special columns for discussing animal electricity. The *Giornale della letteratura italiana* referred to "some truth useful to the service of man." Even Alessandro Volta was convinced, at first, noting that concerning this "miracle" he had passed from "incredulity to fanaticism."[80] The impact on Europe in general was compared to that of the French Revolution itself.

Hopes for understanding the very forces that made life possible, and for realizing human progress through electricity, were not entirely dampened by

Volta's apparent disproof of Galvani's theories. Not only because, as contemporaries recognized, "objections, doubts and disputes, accompanied by wholesome criticism, promote the sciences" rather than destroying them.[81] Giovanni Aldini continued to support the research of his uncle, Galvani, insisting that the voltaic pile actually confirmed the latter's discoveries, by providing chemical causes for the phenomena going on in the muscles. Rather than Leyden jars or capacitors, as Galvani had surmised, perhaps the muscles were in fact nothing but so many voltaic piles, which produced electricity and condensed and conducted it to bring about contractions in the tissues. And even as the voltaic pile was being saluted as a "prodigy of physics, the glory of our century," catalyst of "an entire revolution in favor of Volta's theories,"[82] nonetheless, there was no question that electricity was an important part of muscular movements. Nor could anyone deny that there was much promise in the new joining of physics with physiology. Volta's research, rather than limiting the field, seemed to reveal still more possible applications of technical knowledge to human behavior. Although no immediate use could be found for the pile, the notion was now clearly established that the electrical force could be tamed and manipulated in new ways. Now, in truth, a new era had begun.

To be sure, the new role of science did not go uncontested; and the controversies about Mesmerism, dowsing, and other popular beliefs had some negative effects. Already in 1788, Nicola Valletta reproached academic scientists for their unfounded arrogance in dismissing knowledge achieved by ways they did not approve. Let them remember the Renaissance skeptical tradition, he insisted. In the heat of the debate, he never mentioned that the better-accredited scientists, recalling Galileo's warnings about too much theory and not enough observation, would have been among the first to agree with him that, "human intelligence is not an adequate measure of the truth, and the indisputable verities filling the universe are also incomprehensible." Who among the circle of Spallanzani, Galvani, and Volta could possibly deny that "the infinite marvels of nature are such as to humble once and for all the presumptuous confidence of the human spirit and convince it of its own weakness"? Where they might disagree with Valletta was in attempting to explain the evil eye on the basis of physical influences, putting an ersatz scientific spin on popular convictions that had nothing to do with science.[83]

The attack on science was not, of course, limited to Italy.[84] And if Goethe's impression of knowledge gone awry seemed to hearken back to the traditions of Renaissance magic and Hermeticism, the interpretation of Mary Shelley after the turn of the century pointed instead directly to the brave new world of modern scientism introduced by Galvani. In fact, Dr. Frankenstein, the inventor of the fearsome techno-monster, rejects the very sources upon which Goethe's Faust seemed to rely—Cornelius Agrippa, Albertus Magnus, Paracelsus, and the other occult philosophers of bygone epochs—in favor of "electricity and galvanism."[85] But the creature he fabricates by synthesizing a life force by the modern techniques is born to do evil, not good. To prevent the creation of a race of monsters, Frankenstein refuses the creature's demand for a fabricated bride in return for a promise to desist from rampage; and he is made to suffer the consequences of his refusal. Science, in Shelley's view, is a dangerous obsession; and it may let loose in the world certain products with untold evil effects. Modern ecology movements have raised similar concerns.

In spite of the scientists' best efforts to translate their message into terms that non-scientists could fathom, the growing complexity of scientific work, and its sectorialization into so many specialized professional subdisciplines, increased the potential for incomprehension. The emergence of what would later be called "two cultures"—scientific and nonscientific—led inevitably to misapprehensions and suspicions. Mary Shelley warned about a half-discredited science that she barely understood and whose potential, even in the claims of its most ardent supporters, was nowhere near her worst fears.

One thing was certain: the public sphere of science could no longer be ignored. Never again could the seekers after natural knowledge imagine themselves in a conversation only among a few of their colleagues or with posterity. Indeed, many of the issues regarding this public sphere in the seventeenth and eighteenth centuries remain important features of debates about it to this day. It continues to widen, as new media of communication put more and more technical knowledge into the hands of the non-specialist, raising expectations for what science might be able to accomplish in the future, and raising the stakes for scientists to lay claims as the bearers of that promise.

Notes

1. Carlo Antonio Pilati, *Nuovo progetto di una riforma d'Italia, o sia dei mezzi di liberar l'Italia dalla tirannia dei pregiudizi* . . . 3 vols., 3rd ed. ("Londra": C. Thompson, 1786), 3: 142.

2. The relation of technology to political and social problems is sketched out brilliantly in Emilio Sereni, *History of the Italian Agricultural Landscape*, trans. R. Burr Litchfield (Princeton: Princeton University Press, 1997). In general, concerning the eighteenth-century academies, Eric Cochrane, *Tradition and Enlightenment in the Tuscan Academies* (Chicago: University of Chicago Press, 1961), Brendan Dooley, "Le accademie," *Storia della cultura veneta*, vol. 5: *Il Settecento*, ed. Girolamo Arnaldi and Manlio Pastore Stocchi (Vicenza: Neri Pozza, 1986), 1: 77-90; Michele Lecce, *L'agricoltura veneta nella seconda metà del Settecento* (Verona: n.p., 1958); Paolo Preto, "L'agricoltura bellunese nella seconda metà del Settecento e l'Accademia degli Anistamici," *Critica storica* 15 (1978): 64-107; U. Baroncelli, "L'Accademia agraria di Brescia (secolo XVIII)," *Archivio storico lombardo* 97 (1970): 53. Some information may also be gleaned from Francesco Coletti, *Le associazioni agrarie in Italia dalla metà del secolo decimottavo alla fine del decimonono* (Rome: Unione cooperativa editrice, 1901). For the European context, James E. McClellan, III, *Science Reorganized. Scientific Societies in the Eighteenth Century* (New York: Columbia University Press, 1985).

3. Alexander I. Grab, *La politica del pane. Le riforme annonarie in Lombardia nell'età teresiana e giuseppina* (Milan: Angeli, 1986), 54. In general, Franco Venturi, *Settecento riformatore*, vol. 5: *L'Italia dei lumi (1764-90)* (Turin: Einaudi, 1987), 1: chap. 1.

4. Examples are in Venturi, *Settecento riformatore*, vol. 5: *L'Italia dei lumi (1764-90)*, 300, 360, 370, and passim.
5. Concerning the system in general, Grab, *La politica del pane*, chap. 2, passim.
6. See Jean-Claude Waquet, *Corruption: Ethics and Power in Florence, 1600-1770*, translated by Linda McCall (Cambridge: Polity Press, 1991), concerning the grain scandal of 1748.
7. For instance, Athol Fitzgibbons, *Adam Smith's System of Liberty, Wealth, and Virtue: the Moral and Political Foundations of* The Wealth of Nations (Oxford: Clarendon Press, 1995); William Oliver Coleman, *Rationalism and Anti-Rationalism in the Origins of Economics: the Philosophical Roots of 18th-Century Economic Thought* (Aldershot, Hants, England: E. Elgar, 1995).
8. Venturi, *Settecento riformatore*, vol. 5: *L'Italia dei lumi (1764-90)*, 294, 301, 334.
9. Franco Venturi, *Settecento riformatore*, vol. 5, pt. 2: *La Repubblica di Venezia (1761-1797)* (Turin: Einaudi, 1990), 116, and, in general, Mario Mirri, *La lotta politica in Toscana intorno alle 'riforme annonarie' (1764-75)* (Pisa: Pacini, 1972). Perhaps with some exaggeration, Renato Pasta reads the culture of this period exclusively in terms of the effects of political hegemony, in "Scienza e istituzioni nell'età Leopoldina. Riflessioni e comparazioni," Vieri Becagli and Renato Pasta, eds., *La politica della scienza. Toscana e stati italiani nel tardo Settecento. Atti del Convegno di Firenze, 27-29 gennaio 1994* (Florence: Olschki, 1996), 3-34. I am arguing for a more market-oriented approach than, say, Charles Coulston Gillispie, *Science and Polity in France at the End of the Old Regime* (Princeton: Princeton University Press, 1980). Consider also Jan Golinski, *Science as Public Culture: Chemistry and Enlightenment in Britain, 1760-1820* (Cambridge: Cambridge University Press, 1999), and, for France, Geoffrey V. Sutton, *Science for a Polite Society: Gender, Culture and the Demonstration of Enlightenment* (Boulder, Colo.: Westview Press, 1995), Larry Stewart, *The Rise of Public Science: Rhetoric, Technology and Natural Philosophy in Newtonian Britain, 1660-1750* (Cambridge: Cambridge University Press, 1992); and finally, the essays in William Clark, Jan Golinski, and Simon Schaffer, *The Sciences in Enlightened Europe* (Chicago: University of Chicago Press, 1999).
10. Hanns Gross, *Rome in the Age of Enlightenment. The Post-Tridentine Syndrome and the Ancien Regime* (Cambridge: Cambridge University Press, 1990), 104-08; Baron Karl Otmar von Aretin, "Bernardo Tanucci e il problema dell'assolutismo illuminato nei paesi cattolici," Raffaele Ajello and Mario D'Addio, eds., *Bernardo Tanucci: statista, letterato, giurista: atti del convegno internazionale di studi per il secondo centenario, 1783-1983* (Naples: Jovene, 1986), 457-73.
11. Ildebrando Imberciadori, "L'agricoltura al tempo dei Lorena," Zeffiro Ciuffoletti and Leonardo Rombai, eds., *La Toscana dei Lorena: riforme, territorio, società: atti del convegno di studi (Grosseto, 27-29 novembre 1987)* (Florence: L.S. Olschki, 1989), 139-58; idem, "I moderati toscani e la tradizione leopoldina," Clementina Rotondi, ed., *I Lorena in Toscana: convegno internazionale di studi (Firenze 20-21-22 novembre 1987)* (Florence: Olschki, 1989), 121-38; and Adam Wandruszka, "Pietro Leopoldo," ibid., 31-44. Also see Vieri Becagli, "La pipa di gesso di Pietro Leopoldo," Alessandra Contini and Maria Grazia Parri, eds., *Il Granducato di Toscana e i Lorena nel secolo XVIII: incontro internazionale di studio, Firenze, 22-24 settembre 1994* (Florence: L. S. Olschki, 1999), 285-326; and the larger context noted in Bernardo Sordi, *L'amministrazione illuminata. Riforma delle comunità e progetti di costituzione nella Toscana Leopoldina* (Milan: Giuffrè), 122.
12. Franco Venturi, *Settecento riformatore*, vol. 2: *La chiesa e lo stato dentro i loro limiti* (Turin: Einaudi, 1976), chap. 6. Consider also Marco Santillo, "Brevi note critiche sugli economisti veneti del tardo Settecento," *Clio* 30 (1994): 295-304; and Salvatore Ciriacono, "Agricoltura e agronomia a Venezia e nella Germania del nord: un approccio

comparativo (fine Settecento inizi dell'Ottocento), Roberto Finzi, ed., *Fra studio, politica e economia: la società agraria dalle origine all'età giolittiana. Atti del sesto convegno internazionale, Bologna, 13-15 dicembre, 1990* (Bologna: Istituto per la storia di Bologna, 1992).

13. Renato Pasta, "L'Accademia dei Georgofili e la riforma dell'agricoltura," *Rivista storica italiana* (1993): 484-501; Danilo Barsanti, "Progetti di risanamento della Maremma senese nel secolo XVIII," *Rassegna storica toscana* 25 (1979): 25ff.

14. In his *Ragionamenti sopra l'agricoltura*, cited in Cochrane, *Tradition and Enlightenment*, 151.

15. *Opere di Ludovico Antonio Muratori*, ed. Giorgio Falco and Fiorenzo Forti, 2 vols. (Milan: Ricciardi, 1960), 2: 1579. Concerning this work, Franco Venturi, *Settecento riformatore*, vol. 1: *Da Muratori a Beccaria* (Turin: Einaudi, 1969), chap. 2.

16. *Novelle letterarie* 2 (1741): 760: "There is no need to blame the philosopher who pays attention to the precepts of the illiberal arts—who concerns himself with the instruments and operations and labor of artisans; and studies these operations in books about them. . . . It is likewise necessary for him to have his hands ready, industrious and agile for handling instruments of those arts, so his experiments will be more exact."

17. *Novelle letterarie* 12 (1751): 542; 1 (1740): 429.

18. *Illuministi italiani*, vol. 5: *Riformatori napoletani*, ed. Franco Venturi (Turin: Einaudi, 1962), 269, from his *Logica per giovanetti*. In general, Eluggero Pii *Antonio Genovesi dalla politica economica alla politica "civile"* (Florence: Olschki, 1984).

19. *L'agricoltore sperimentato* (Naples: 1769); *Elementa physicae* (Naples: 1745), in collaboration with the physicist Giuseppe Orlandi.

20. *Illuministi italiani*, vol. 5: *Riformatori napoletani*, 86, 120, from *Il vero fine delle lettere e delle scienze* (1753).

21. On this whole problem see also Bianca Fadda, *L'innesto del vaiolo: un dibattito scientifico e culturale nell'Italia del Settecento* (Milan: Angeli, 1987).

22. *Atti dei fisiocritici* 1 (1760), entire. In addition, Giovanni Targioni Tozzetti, *Relazioni d'innesti di vajuolo fatto in Firenze nell'autunno dell'anno 1757* (Florence: Bonducci, 1757). Note that Condamine's work was translated as *Memoria sull'inoculazione del vajuolo* and published in the *Magazzino toscano* 2 (1755): 74-81, 145-62, 193-204.

23. Gemma Prontera, "Medici, medicina e riforme nella Firenze del Settecento," *Società e storia* 26 (1984), 793.

24. Quoted in Michele Lecce, *L'agricoltura veneta*, 29. See Giampiero Fumi, "Pietro Arduino" and "Giovanni Arduino" in S. Zaninelli, ed., *Scritti teorici e tecnici di agricoltura*, vol. 2: *Dal Settecento agli inizi dell'Ottocento* (Milan: Il Politilo, 1989), 107-64.

25. Ezio Vaccari, *Giovanni Arduino (1714-1795). Il contributo di uno scienziato veneto al dibattito settecentesco sulle scienze della Terra* (Florence: Olschki, 1993), chap. 4.

26. Padova, Accademia Patavina di Scienza, Lettere ed Arti, Giornale C, fols. 263r-285v; Venice, Museo Correr, mss. P.D. 747c: "Elenco dei documenti riguardanti l'Accademia di Agricoltura, Commercio e Arti negli Antichi Archivi Veronesi," containing a list of projects begun and prizes awarded; Vicenza, Biblioteca Bertoliana, Accademia, 12, circulars for 6 March 1771 concerning prizes, 16 May 1771 concerning prizes, 3 June 1771 concerning Venetian government queries, 16 August 1771 concerning experiments, and so forth. Consider also *Raccolta di memorie delle pubbliche accademie di agricoltura, arti e commercio dello stato veneto*, 18 vols. (Venice: G. A. Perlini, 1789-97), vol. 1, passim.

27. Venice, Biblioteca Nazionale Marciana, mss. ital. VII: 1951 (=8833), no. 28: Pietro Arduino, "Piano per l'Istituzione dell'Accademia di Agricoltura di Padova," 2 May 1769.

28. *Estratto del trattato dell'utilità morale, economica e politica delle accademie d'agricoltura, arti e commercio,* in Pietro Custodi, ed., *Scrittori classici italiani di economia politica. Parte moderna,* vol. 19 (Milan: Destefanis, 1806), 140, 395.

29. *Elementi di economia pubblica* in Pietro Custodi, ed., *scrittori italiani di economia politica, parte moderna,* vol. 11 (Milan: Destefanis, 1804), 153. Also, 22: "Neither the products of the earth, nor the operations of the hand, nor commerce, nor public tributes [the bases of economics and public happiness], can ever reach perfection among men if men do not know the moral and physical laws concerning the things upon which they act; if the development of bodies is not accompanied by the development of social habits; if among the multitude of individuals, labor and products are not always subject to some sort of order that makes all operations easy and secure." Consider the articles by Pier Luigi Porta and Angelo Moioli in *Cesare Beccaria tra Milano e l'Europa. Convegno di studi per il 250º anniversario della nascita* (Milan-Bari: Cariplo-Laterza, 1990), 329-70.

30. Luigi dal Pane, *Lo Stato pontificio e il movimento riformatore del Settecento* (Milan: Giuffré), 714.

31. *Saggi scientifici e letterari dell'Accademia di Padova* (Padua: 1786-94), 2: v.

32. Ibid., xlix, xl.

33. *Pensieri sulla pubblica felicità* (Rome: Casaletti, 1774), cited in *Novelle letterarie* 37, n.s. 6 (1775): 90.

34. Letter to Francesco Griselini, 1764, in *Illuministi italiani,* vol. 7: *Riformatori delle antiche repubbliche, dei ducati, dello stato pontificio e isole,* ed. Giuseppe Giarrizzo, Gianfranco Torcellan and Franco Venturi Turin: Einaudi, 1965), 104.

35. Zanon, *Estratto del trattato dell'utilità morale, economica e politica delle accademie d'agricoltura, arti e commercio,* 145.

36. Exclusively political aspects are discussed in Vieri Becagli, "Economia e politica del sapere nelle riforme leopoldine delle accademie," Vieri Becagli and Renato Pasta, eds., *La politica della scienza,* 35-66. Compare, for a later period, Barbara Maffiodo, "L'Accademia delle Scienze di Torino e la promozione della medicina in Piemonte," Ibid., 319-44; and Pasta, *Scienza politica rivoluzione. Lópera di Giovanni Fabbroni, 1752-8122, intellettuale e funzionario al servizio dei Lorena* (Florence: Olschki, 1989).

37. *Magazzino toscano* 1 (1754), preface. Compare *Giornale d'Italia spettante alla scienza naturale e principalmente all'agricoltura e al commercio* 1 (1764): 65, 68, 124, 167, 416; 2 (1765): 57, 78.

38. Cited in Gianfranco Torcellan, "Un problema aperto: Politica e cultura nella Venezia del Settecento," *Settecento veneto e altri scritti* (Turin: Giappichelli, 1969), 314. Concerning Scottoni, see Mario Infelise, "Appunti su Giovanni Francesco Scottoni, illuminista veneto," *Archivio veneto,* ser. 5, vol. 119 (1982): 39-73. The late eighteenth-century periodicals are well analyzed in Giuseppe Ricuperati, "Giornali e società nell'Italia delle riforme," *La stampa italiana dal Cinqecento all'Ottocento* ed. Valerio Castronovo and Nicola Tranfaglia, 2nd ed. (Bari: Laterza, 1986), 191-353.

39. Compare Georges Gusdorf, *L'avènement des sciences humaines au siècle des lumières* (Paris: Payot, 1973), as well as articles by Roy Porter, Sylvana Tomaselli and David Carrithers in *Inventing Human science: Eighteenth-Century Domains,* ed. Christopher Fox, Roy Porter, and Robert Wokler (Berkeley: University of California Press, 1995).

40. Concerning Targioni Tozzetti's career, Tiziano Arrigoni, *Uno scienziato nella Toscana del Settecento: Giovanni Targioni Tozzetti* (Florence: Gonnelli, 1987). In addition, see Ezio Vaccari, "Cultura scientifico-naturalistica ed esplorazione del territorio: Giovanni Arduino e Giovanni Targioni Tozzetti," Giulio Barsanti, Vieri Becagli and Renato Pasta, eds., *La politica della scienza,* 243-64; Giuseppe Guerrini, "Le scienze al tempo dei Lorena e l'opera di Giovanni Targioni Tozzetti," Zeffiro Ciuffoletti and Leonardo Rombai, eds., *La Toscana dei Lorena,* 361-78.

41. *Relazione delle febbri che si sono provate epidemiche in diverse parti della Toscana l'anno 1767, scritta per ordine del'illustriss. e chiar. Magistrato di sanità* (Florence: Cambiagi, 1767)

42. *Vera natura, cause e tristi effetti della ruggine*, ed. Gabriele Goidanich (Rome: Accademia di Italia, 1943)

43. *Breve istruzione circ'ai modi di accrescere il pane col mescuglio d'alcune sostanze vegetabili alle quali si sono aggiunte certe nuove e più sicure regole per ben sceglier i semi del grano da seminarsi nel corrente autunno del 1766* (np. nd.).

44. *Viaggio in Dalmazia* (Venice: Milocco, 1774). On this, see also Franco Venturi, *Settecento riformatore*, vol. 5, pt. 2: *La Repubblica di Venezia (1761-1797)*, 81.

45. Luca Ciancio, *Autopsie della terra. Illuminismo e geologia in Alberto Fortis, 1741-1803* (Florence: Olschki, 1995), 174-84.

46. On the relation between Vallisneri senior and his son, see Simone Contardi, *La rivincita dei 'filosofi di carta.' Saggio sulla filosofia naturale di Antonio Vallisneri Junior* (Florence: Olschki, 1994).

47. Ciancio, *Autopsie della terra*, 197-217. Compare Vincenzo Ferrone, *I profeti dell'illuminismo. Le metamorfosi della ragione nel tardo Settecento italiano* (Bari: Laterza, 1989), chap. 3.

48. On this topic, John L. Heilbron, *Electricity in the Seventeenth and Eighteenth Centuries. A Study of Early Modern Physics* (Berkeley: University of California Press, 1979), chap. 15; and S. Ramazzotti and Luigi Briatore, "Appunti di storia della fisica. Dalle calze di seta di Symner all'elettroforo di Volta," *Giornale di fisica* 15 (1974): 52-59.

49. *Nuovo dizionario scientifico e curioso, sacro-profano*, 10 vols. (Venice: Milocco, 1746-51), 6: 953. Some information on Pivati is in Silvano Garofalo, *L'enciclopedismo italiano, Gianfrancesco Pivati* (Ravenna: Longo, 1980).

50. Ciancio, *Autopsie della terra*, 217-30.

51. Daniela Silvestri, "Lazzaro Spallanzani e Alberto Fortis," Giuseppe Montalenti and Paolo Rossi, eds., *Lazzaro Spallanzani e la biologia del Settecento. Teorie, esperimenti, istituzioni scientifiche* (Florence: Olschki, 1982), 315. A useful bibliographical compendium concerning Spallanzani is Claude E. Dolman's entry in the *Dictionary of Scientific Biography* 12 (1975): 553-67.

52. Walter Bernardi, *I Fluidi della vita. Alle origini della controversia sull'elettricità animale* (Florence: Olschki, 1992), 170-71, citing *Edizione nazionale delle Opere di Lazzaro Spallanzani. Parte prima. Carteggi*, 12 vols., ed. Pericle Di Pietro (Modena: Mucchi, 1984-90), 3: 118. The earlier work in question was *Opuscoli di fisica animale e vegetabile*, 2 vols. (Modena: 1776).

53. But see Carlo Castellani, "Lazzaro Spallanzani nei suoi rapporti con la scienza e la cultura del Settecento," Giuseppe Montalenti and Paolo Rossi, eds., *Lazzaro Spallanzani e la biologia del Settecento*, 21-44. The many facets of Spallanzani's career are examined in this anthology, in articles by François Duchesneau, 45-66; Thomas S. Hall, 67-83; R. Milani, 83-108; Ferdinando Abbri, 121-36, Massimo Aloisi, 137-55; and others mentioned below.

54. *Dissertazioni di fisica animale e vegetale*, 2 vols. (Modena: 1780), translated as *Dissertations Relative to the Natural History of Animals and Vegetables*, 2 vols. (London: 1784), vol. 1, part 1, dissertation 5. The dissertations on digestion were also published separately as *Experiences sur la digestion de l'homme et de différents espèces d'animaux....* (Geneva: 1785).

55. Consider John Farley, *The Spontaneous Generation Controversy from Descartes to Oparin* (Baltimore: Johns Hopkins University Press, 1977); Renato G. Mazzolini and Shirley A. Roe, eds., *Science Against the Unbelievers: the Correspondence of Bonnet and Needham, 1760-1780* (Oxford: Voltaire Foundation at the Taylor Institution, 1986);

Science and the Marketplace   159

Shirley A. Roe, *Matter, Life, and Generation: Eighteenth-Century Embryology and the Haller-Wolff Debate* (Cambridge: Cambridge University Press, 1981); in addition, articles in Giuseppe Montalenti and Paolo Rossi, eds., *Lazzaro Spallanzani e la biologia del Settecento*, by G. Pancaldi, 283-94; Shirley. A. Roe, 295-304; R. Toellner, 109-20.

56. Ferdinando Abbri and Walter Bernardi, "Introduzione," Fabrizia Capuana and Paola Manzini, eds., *La mal-aria di Lazzaro. Spallanzani e la respirabilità dell'aria nel Settecento* (Florence: Olschki, 1996), vii.

57. Paola Manzini, "Il manoscritto ritrovato," in F. Capuana and P. Manzini, eds., *La mal-aria di Lazzaro. Spallanzani.*

58. Concerning which, Alfredo Jona, *La collezione monumentale di Lazzaro Spallanzani classificata e ordinata secondo lo stato della scienza alla fine del secolo XVIII*, 2nd ed. (Reggio Emilia: Tipo-Litografia degli Artigianelli, 1959).

59. Ezio Vaccari, "Contributo di Spallanzani alle scienze della terra del Settecento," Walter Bernardi and Paola Manzani, eds., *Il cerchio della vita. Materiali di ricerca del Centro Studi Lazzaro Spallanzani di Scandiano sulla storia della scienza nel Settecento* (Florence: Olschki, 1999), 138.

60. *Viaggi alle Due Sicilie e in alcune parti dell'Appenino* (Milan: Classici italiani, 1825), 21.

61. Marta Cavazza, "Laura Bassi e il suo gabinetto di fisica sperimentale: realtà e mito," *Nuncius* 10 (1995): 724-26; idem, "Laura Bassi 'maestra' di Lazzaro Spallanzani," W. Bernardi and P. Manzani, eds., *Il cerchio della vita*, 185-202. Consider also, Gabriella Berti Logan, "The Desire to Contribute: An Eighteenth-Century Woman of Science," *American Historical Review* 99 (1994): 785ff, and, regarding the patronage aspects of Bassi's career, Paula Findlen, "Science as a Career in Enlightenment Italy. The Strategies of Laura Bassi," *Isis* 84 (1993): 441-69.

62. *De lapidibus ab aqua resilientibus* (Bologna: Soliani, 1765).

63. Ottavio Barnabei, "Some Information about the Academy of Sciences," in *Discorsi e scritti in onore di Luigi Galvani nel bicentenario della morte, 1798-1998* (Bologna: Forni, 1999), 15. In addition, Marta Cavazza, *Settecento inquieto. Alle origini dell'Istituto delle Scienze di Bologna* (Bologna: Il Mulino, 1990); as well as the exhibition catalogue, *I materiali dell'Istituto delle Scienze* (Bologna: Accademia delle scienze, 1979).

64. All are examples from the Institute's *Commentarii* 3 (1755), 31, 52, 90, 80, 94, 132.

65. Leonardo Giardina, *Lezioni inedite di ostetrica di Luigi Galvani* (Bologna: STEB, 1965), xxxv.

66. John Heilbron, "The Contributions of Bologna to Galvanism," *Historical Studies in the Philosophical and Biological Sciences* 22 (1991): 61.

67. *Sentimento del dott. Luigi Galvani sopra la natura del male da cui sono attaccate le bestie bovine nelle comunità di Vimignano e Savignano, di Vigo e Verzuno, di Burzanella e di Montagnragazza, e di Camugnano* (1775, repr. Bologna: La Nuova Veterinaria, 1937).

68. Marco Bresadola, "La biblioteca di Luigi Galvani," *Annali della storia delle università italiane* 1 (1997): 167-97.

69. Giuseppe Veratti, *Osservazioni fisico-mediche intorno alla elettricità* (Bologna: Dalla Volpe, 1748), 139.

70. J. Heilbron, *Electricity in the 17th and 18th Centuries*, 354.

71. Cited in Marco Bresadola, "Medicina e elettricità," in *Discorsi e scritti in onore di Luigi Galvani*, 55-64.

72. *De ossibus lectiones quattuor* (Bologna: Compositori, 1966), quoted in Marco Piccolini, "Luigi Galvini e l'elettricità animale," in *Discorsi e scritti in onore di Luigi Galvani*, 26. Concerning what follows, Marcello Pera, *La rana ambigua. La controversia*

sull'elettricità animale tra Galvani e Volta (Turin: Einaudi, 1986), translated as *The ambiguous frog : the Galvani -Volta controversy on animal electricity*, trans. Jonathan Mandelbaum (Princeton, N.J.: Princeton University Press, 1992), chaps. 1-3; Walter Bernardi, *I fluidi della vita*, part 1. I lean toward Bernardi's interpretation of the crux of the controversy.

73. Carlo Capra, *Giovanni Ristori, da illuminista a funzionario* (Florence: La Nuova Italia, 1968), chap. 2.

74. C. Capra, *Giovanni Ristori,* 96.

75. Quoted in Martino Capucci and Renzo Cremante, "Aspetti della circolazione delle idee nelle *Memorie enciclopediche* (1781-87)," *Produzione e circolazione libraria a Bologna nel Settecento. Avvio di un'indagine. Atti del V colloquio, Bologna, 22-23 febbraio, 1985* (Bologna: Istituto per la storia di Bologna, 1987), 362.

76. On this period, Carlo Zaghi, *L'Italia di Napoleone dalla Cisalpina al Regno* (Turin: Utet, 1986); Edoardo Tortarolo, *Illuminismo e rivoluzioni. Biografia politica di Filippo Mazzei* (Milan: Angeli, 1986); Ivan Tognarini, ed., *La Toscana nell'età rivoluzionaria e napoleonica* (Naples: Edizioni scientifiche, 1985).

77. Giovanni Aldini, *Memoria intorno all'elettricità animale* (Bologna: np, 1794), 1. Compare Eusebio Valli, *Experiments on Animal Electricity with their Applications to Physiology* (London: J. Johnson, 1794), 177ff; Richard Fowler, *Experiments and Observations Relative to the Influence Lately Discovered by M. Galvani and Commonly Called Animal Electricity* (Edinburgh: Duncan, 1793).

78. *The Effects of Animal Electricity on Muscular Motion*, ed. I.B. Cohen, 82-88. Consider Paolo Rossi, "Sulle origini dell'idea di progresso," Evandro Agazzi, ed., *Il concetto di progresso nella scienza* (Bologna: Il Mulino, 1976), 37-88.

79. Gioacchino Carradori, *Istituzione del galvanismo in Italia, o sia della contesa fra Volta e Galvani* (Florence: all'insegna dell'Ancora, 1817), 33.

80. Citations in W. Bernardi, *I fluidi della vita*, 52-53.

81. G. Carradori, *Istituzione*, 47.

82. G. Carradori, *Istituzione*, 42.

83. I used Nicola Valletta, *Cicalata sul fascino volgarmente detto jettatura* (Milan: Alberto Fidi, 1925). Concerning this work and the issues connected with it, Vincenzo Ferrone, *I profeti dell'illuminismo*, chap. 8. Compare Robert Darnton, *Mesmerism and the End of the Enlightenment in France* (Cambridge, Mass.: Harvard University Press, 1968); Geoffrey V. Sutton, "Electric Medicine and Mesmerism," *Isis* 72 (1981): 375-92; Patricia Fara, *Sympathetic Attractions: Magnetic Practices, Beliefs and Symbolism in Eighteenth-Century England* (Princeton: Princeton University Press, 1996); and Luca Ciancio, *Autopsie della terra*, 217-18.

84. In general, see Francesco Rigotti, *L'umana perfezione. Saggio sulla circolazione e diffusione dell'idea di progresso nell'Italia del primo Ottocento* (Naples: Bibliopolis, 1980); also, Marina Caffiero, *La nuova era: miti e profezie dell'Italia in Rivoluzione* (Genova : Marietti, 1991).

85. Mary Shelley, *Frankenstein, or The Modern Prometheus*, ed. with an Introduction by M. K. Joseph (London: Oxford University Press, 1969), 41.

Epilogue

So far we have considered the impact of a marketplace on science. For a moment, let us now consider the reverse—i.e., the impact of science on capitalism before the Industrial Revolution. We must keep in mind that the marketplace in the early modern period was not the marketplace of modern capitalism, whose seeds were nonetheless being sown. The structures of commerce did not extend into every nook and cranny of existence. Nor was the science we have analyzed entirely identifiable with its modern counterpart, a social institution of enormous power and authority. However, we would be just as mistaken to undervalue the connection between the production of natural knowledge and the production of other commodities, as we would be to underestimate the importance of the culture of that knowledge in the early modern world. And there can be no doubt that the development of knowledge about nature helped generate new products—at the very least, new products of the printing press; and it helped sell books. As far as the economic impact of the ideas themselves in this period is concerned, our answer will have to be less positive and far more hypothetical.[1]

To be sure, the most influential ideas about nature in this period were not dispensed by the famous professors in Italian universities. Nor were they dispensed necessarily by learned amateurs attempting to stay abreast of pan-European debates in the fields of natural knowledge. They came from popular books of medical secrets and farmers' almanacs. The latter constituted a busy tertiary sector serving families and their productive activities through the end of the eighteenth century. Along with information about herbal medicines and magical invocations, they supplied readers with a wealth of wisdom gleaned from the most diverse sources, such as (in the words of the late sixteenth-century Neapolitan *Almanacco perpetuo*), "do not plant grain in wet ground" or "do not plant trees in new furrows."[2] On the basis of astrological principles, they gave information about what years would bring good crops and what years would bring excessive rainfall. Inevitably, ideas deriving from the high culture of science seeped in, as when the *Palmaverde*, published in mid-eighteenth-century Turin, acknowledged that "modern observers" no longer held the heavens to be immutable or the number of heavenly bodies to be finite.[3] Although the circulation of such information is not known to have contributed measurably to agricultural productivity, it may well have provided some slight impulse to literacy in the countryside. And if Gregorio Leti's observation of Bologna farmers tilling the soil with a book in one hand is surely an exaggeration, nonetheless, the habit of regarding agriculture as a matter of intellectual disquisition set the stage for the full emergence of agronomy as a science in the eighteenth century.[4]

The closer the encyclopedic and specialized journals came to the high culture of science, the more remote they may have seemed from the realm of production. Yet they served a tertiary sector of public health experts for whom articles on medicines, therapies, anatomical discoveries, mineral springs, botany, and others found in a typical volume of the *Giornale de' letterati* published in Rome in the late seventeenth century, were no doubt just as compelling as news about a large protuberance on the right hand of a girl from Toulouse. Articles on navigation or on the tides appear to have more relevance to productivity than do others on conic sections; although higher mathematics and physics were important foundations on which to build intellectual infrastructure. By the time of the mid-eighteenth-century heyday of the French *Encyclopédie* and its Italian offshoots, economics and agronomy were among the new disciplines regularly included in periodicals such as the Venice-based *Giornale d'Italia* and the Livorno-based *Magazzino toscano*.

Numerous inventions and technologies from this period had a decisive impact on the economy. In the particular field of clock making, where time management was born and bred, we may cite the pendulum, a product of the investigations of Galileo and Huygens.[5] In the field of ballistics, where the basic principles of engineering were being forged, along with a thriving industry in firearms, key contributions were made by mathematicians from Niccolò Tartaglia to Galileo and beyond. Perhaps the best example of all was the field of hydraulics. In Italy, as in Holland, theoretical concepts elaborated by natural philosophers were regularly applied to practical projects for land reclamation and irrigation aimed at agricultural development; and we have already seen that representatives of the Galilean school were recruited by governments for this purpose. Grand Duke Cosimo III of Tuscany sent Pietro Guerrini on a hydraulics fact-finding mission throughout Holland, Flanders, France, Germany and England in the 1680s. It would be interesting to see what contacts Gillis van den Houte, Egidio Vandenhout and Willem de Raet, northern European experts on water, had with the world of the natural philosophers in Italy, where they worked.[6]

That the development of useful science in Italy was at least as often driven by political as by economic motivations made no more difference in Italy than it did in France. The Venetian government's interest in the telescope for spotting enemy ships did not detract from the device's utility for other purposes. By the same token, that Venice and other states sponsored land reclamation projects mainly in order to extend the ruling elites' territorial reach did not make the new methods any less effective for increasing arable space. Ballistics, meanwhile, was the classic wartime science with peacetime uses—cultivated with equal ardor at the Turin Royal School of Artillery in the time of Louis Lagrange and the Verona Military School in the time of Anton Maria Lorgna.[7]

What made possible the joining of artisinal culture with scientific culture in the eighteenth century was a tradition of borrowings between the two sectors, dating at least from a century before.[8] And while machines like the ones discussed by Agostino Ramelli, in the tradition of Leonardo da Vinci, came out of the shop of an artisan, Galileo's observations on water in motion were ultimately as important as Torricelli's discoveries concerning air pressure for the development of the steam engine. That the engine was finally built by an artisan, Thomas Newcomen, working in England, was a consequence of the shared culture of practical science as well as of basic economic and social differences between England and Italy that we do not have space to pursue here in great detail.

Variations in the dynamics of capitalism between the north and south of Europe (and between the north and south of Italy) on the verge of the Industrial Revolution had as much to do with the diffusion of natural resources as with the prevailing forms of government; they were as much affected by demographics as by the spread of literacy. The agrarian class structure discussed by Robert Brenner was no doubt at least as influential as the centralization of justice systems discussed by Alan Macfarlane. And inequalities both in scientific culture and economic culture were symptoms of yet other cultural and social differences, some of them discussed by Weber and Tawney and recently reproposed with a few modifications by David Landes. The most misleading explanation is a monocausal one.[9]

In Italy, rather than a direct impact of scientific ideas on the marketplace, we can most accurately discern the operation of a law of unintended consequences. In spite of the exigencies of the philosophers and of the proprietary classes, and in spite of the traumatic military events between the eighteenth and nineteenth centuries that tore the peninsula apart only to forge it together in new ways, the spread of medical knowledge no doubt helped set the stage for the continuous demographic growth that was to characterize the decades to come. Although the results were by no means immediate, agronomical knowledge ensured that this population growth was not wiped out in a vast Malthusian crisis, as occurred repeatedly in the sixteenth and seventeenth centuries. Instead, the premises were laid for the agricultural capital accumulation that would eventually make possible the employment of increased numbers of laborers in industrial operations in Piedmont and Lombardy.

Even more important, by force of the voluminous eighteenth-century press concerning useful knowledge, a general cultural tendency emerged that may have had a wide impact later on—what we might call a scientific ethic in harmony with the spirit of capitalism. This scientific ethic might be viewed as having been composed of three basic elements, whose growing presence in the vocabulary of the time indicated a more and more pronounced affinity. First, we note a critical attitude toward received traditions, even when barely understood. The "freedom" of the thinker was now measured by how much he was "unshackled" to conventional views; "ancient" became a term of disparagement. Next, we note an attitude directed toward the practical, in harmony with commercial productivity—that is, formed of the practice of experiments, the verification of data, the manipulation of natural objects, including those connected with the trades. "Truth" was never so verifiable as when it savored of the "soot of the furnace" or was apprehended by the "sweat" of the researcher. "Mechanic" was no longer necessarily an object of scorn. Finally, we note an attitude open to change, to accepting a world in constant motion. "New" appeared ever more desirable than "old," the "future" ever better than the "past." Transformations in ideas as well as in economic practices seemed to favor growth, improvement, and what we might call, with all the proper reservations, and taking into account the profound differences between one geographical area and another, progress.

Meanwhile, as we have seen, the net effect of a public discussion about the wonders of nature, in a time of institutional reform and curiosity about the possibilities of the application of science, was a refocusing of research goals in new and fruitful directions capable of inspiring new generations of researchers and adding more luster to the reputation of the science practitioner as the savior of humanity, bearer of progress, prophet of improvement. The pattern these practitioners set in their relations to their public and in the direction of their work continued well into the following century, even while Italian contributions for a time were once again eclipsed.

Viewed from this standpoint, the Italian case, with all of its complexity and individuality, seems to follow a common European pattern—the pattern that characterizes the modernity we now share. As science and the marketplace play ever-greater roles in society, an accurate understanding of their relationships becomes increasingly important. We may wonder whether there will ever be a science, or a marketplace, fundamentally different from the ones we have, perhaps more inclusive, multicultural, multiethnic, democratic, poststructuralist, and post-modern.[10] And especially among those who believe science is mainly a reflection of society, such questions are now being hotly debated. For answering such questions, however, the competence of the historian is perhaps less relevant than the imagination of the poet or novelist.

Notes

1. For the following, I am indebted, among other works, to Salvatore Ciriacono, *La Rivoluzione industriale* (Milan: Mondadori, 2000).

2. Rutilio Benincasa, *Almanacco perpetuo* (Venice: Giunti, 1622), 187-88. For what follows, also 116.

3. Lodovica Braida, *Le guide del tempo: produzione, contenuti e forme degli almanacchi piemontesi nel Settecento* (Torino: Deputazione subalpina di storia patria, 1989), 154; in addition, Elena Gremigni, *Periodici e almanacchi livornesi, secoli 17-18* (San Benedetto [Livorno]: 1996); Giuseppe Baretta, Grazia Maria Griffini Rosnati, *Almanacchi milanesi del '700* (Milan: Biblioteca nazionale Braidense, 1996).

4. Leti's comment is in *L'Italia regnante*, 4 vols. ("Valenza" [=Venice]: Guerini, 1675-6), 2: 81.

5. David S. Landes, *Revolution in Time* (Cambridge University Press, 1983).

6. Guerrini's letters to Cosimo III are in ASF, *Archivio Mediceo del Principato*, filza 6390. Thanks to Francesco Martelli for alerting me to them. Concerning the problems

mentioned in this paragraph, Salvatore Ciriacono, *Acque e agricoltura. Venezia, l'Olanda e la bonifica europea in età moderna* (Milan: Angeli, 1994).

7. Vincenzo Ferrone, *La nuova Atlantide e i lumi: scienza e politica nel Piemonte di Vittorio Amedeo III* (Turin: A. Meynier, 1988); Franco Piva, *Anton Maria Lorgna e l'Europa* (Verona: Accademia di Agricoltura, Scienze e Lettere, 1993). Compare Marc Bloch, *Land and Work in Mediaeval Europe*, trans. J. E. Anderson (Berkeley: University of California Press, 1967); and Charles Coulston Gillispie, *Science and Polity in France at the End of the Old Regime* (Princeton: Princeton University Press, 1980).

8. Compare Fernand Braudel, *Civilization and Capitalism, 15th-18th Century*, vol. 1: *The Structures of Everyday Life: the Limits of the Possible*; trans. revised by Sian Reynolds (London: Collins, 1981); as well as James E. McClellan III, *Science and Technology in World History* (Baltimore: The Johns Hopkins University Press, 1999); D. S. L. Cardwell, ed., *The Norton History of Technology* (New York: Norton, 1994), all of whom assert the structural inconsistency of artisinal culture and scientific culture.

9. David S. Landes, *The Wealth and Poverty of Nations: Why Some are so Rich and Some are so Poor* (New York: Norton, 1998); Max Weber, *The Protestant Ethic and the Spirit of Capitalism*, trans. Talcott Parsons (New York: Charles Scribner's Sons, 1958); R. H. Tawney, *Religion and the Rise of Capitalism: a Historical Study* (London: J. Murray, 1926); Robert Brenner "Agrarian Class Structure and Economic Development in Pre-industrial Europe," *Past and Present* 70 (1976): 30-75; Alan Macfarlane, *The Origins of English Individualism: the Family, Property and Social Transition* (Oxford: Blackwell, 1978).

10. For instance, Mario Biagioli, ed., *The Science Studies Reader* (New York: Routledge, 1999).

Index

Abbri, Ferdinando, 158, 159
Abenragel, 22, 36
Accolti, Pietro, 83
Acerbi, Antonio, 137
Adam, Charles, 58
Adelmann, Howard B., 57, 79, 84, 103, 104
Agazzi, Evandro, 160
Ago, Renata, 35, 37
Agrippa, Cornelius, 153
Ajello, Raffaele, 155
Alberigo, Pietro Paolo, 54
Albrizzi, Girolamo, 50, 53, 55, 60
Albumasar, 28
Alciato, Andrea, 67, 69
Aldini, Giovanni, 151, 153, 160
Aldrovandi, Ulisse, 5, 14, 47
Algarotti, Francesco, 113, 124, 130
Allegra, Luciano, 36
Aloisi, Massimo, 158
Altieri Biagi, Maria Luisa, 13, 16
Altobelli, Ilario, 24, 38
Amabile, Luigi, 39, 40
Amaduzzi, Giovanni Cristofano, 14
Ameyden, Theodore, 40
Andrea, Francesco d', 52, 75
Angeli, Stefano degli, 58, 96, 108
Appadurai, Arjun, 35
Appolis, Émile, 130
Aragona, Giovanni d', 82
Aratus Solensis, 22, 36
Arduino, Giovanni, 142, 145, 156
Arduino, Pietro, 142, 156
Aretin, Baron Karl Otmar von, 155
Argoli, Andrea, 25, 36, 96, 108
Aristarchus, 22, 36
Aristotle, 6, 8, 11, 26, 64, 66, 54, 76, 88, 89
Armocida, Giuseppe, 86
Arnaldi, Girolamo, 154
Arrighetti, Niccolò, 115, 124
Arrighi, Gino, 58, 131, 134
Arrigoni, Tiziano, 157
Artigny, Antoine Gachet d', 120

Astorini, Elia, 51, 122
Auzout, Adrien, 43
Ávalos de Aquino, Alfonso II d', (Marqués de Vasto), 80
Avicenna (Ibn-Sina,Ali), 64, 65, 76, 92, 105
Avogadro, Amedeo, 78

Bacchini, Benedetto, 51
Bacon, Francis, 6, 25, 29, 56, 77, 95
Badelli, Antonio, 19
Baglivi, Giorgio, 49, 60, 65
Baillie, Granville Hugh, 133
Baldini, Ugo, xvi, 59, 79, 104, 107, 111, 134
Balsamo, Luigi, 130
Balsimelli, Francesco, 38
Banks, John, 101
Barbarisi, Gennaro, 85
Barbaro, Daniele, 5, 15
Barberini, Francesco, 11
Barberini, Taddeo, 33
Barbieri, Lorenzo, 125
Barbieri, Ludovico, 136
Baretta, Giuseppe, 164
Barnabei, Ottavio, 159
Barnes, Robin Bruce, 37
Baroncelli, Ugo, 154
Baroncini, Gabriele, 59, 80, 134
Barozzi, Nicolò, 39
Barsanti, Danilo, 131, 156
Barsanti, Giulio, 157
Bartoli, Sebastiano, 51, 122
Barzazi, Antonella, 36
Barzman, Karen-edis, 82
Bassi, Laura, 148, 149, 159
Bastiaanse, Alexandro, 40
Beaulieu, A., 57
Becagli, Vieri, 155, 157
Beccaria, Cesare, 112, 142, 157
Beccaria, Giambattista, 149
Bella, Saverio di, 14, 81
Bellarmino, Roberto, 1, 2
Bellini, Lorenzo, 45, 48, 49, 61, 64, 70,

Index

79, 91, 93
Belloni, Luigi, 134
Benedict XIV (Prospero Lambertini), 125, 141
Benincasa, Rutilio, 164
Bentivoglio, Guido, 30
Benvenuti, Giuseppe, 116
Berchet, Guglielmo, 39
Berengo, Marino, 129, 130, 137
Bergia, Silvio, 107
Bérigard, Claude, 46, 59, 88
Bernardi, Walter, 61, 134, 158, 160
Bernoulli, Daniel, 97, 102, 108
Bernoulli, Jacob, 99, 109
Bernoulli, Nicolas, 70
Berthoud, Ferdinand, 133
Berti, Gianlorenzo, 126
Bertini, Giuseppe Maria Saverio, 122
Bertolotti, Antonio, 35
Betto, Bianca, 79, 80
Biagioli, Mario, 13, 39, 165
Bianchini, Francesco, 56, 62
Bilio, Luigi, 37
Bina, Andrea, 124
Bloch, Marc, 165
Bloch, Olivier, 108
Boccone, Paolo, 48, 50, 59, 60, 121
Boehm, Laetitia, 16, 82, 85
Boerhaave, Hermann, 122
Boley, Bruno A., 107
Böll, Franz, 37
Bombardini, Antonio, 102
Bonelli, Benedetto, 137
Bonnet, Charles, 77
Borelli, Giovanni Alfonso, 41, 44, 51, 53, 64, 70, 84, 91, 97, 98, 121, 122
Borghese, Marcantonio, 84
Borghese, Scipione, 42, 59
Borgondio, Orazio, 116, 131
Borsetti, Ferrante, 82
Bortolotti, Ettore, 84, 85
Boscovich, Ruggero Giuseppe, 113, 123, 131, 135
Bose, Georg Matthias, 100
Bossi, M., 132
Bottasso, Enzo, 36
Bouché-Leclerq, Auguste, 38-40
Bourguet, Louis, 121
Bowden, Mary Ellen, 35, 39
Boyer, Carl B., 16
Boyle, Robert, 54, 56, 89, 95, 98, 107

Bracciolini, Francesco, 32
Brahe, Tycho, 26, 47, 95
Braida, Lodovica, 164
Brambilla, Elena, 86
Braudel, Fernand, 165
Brenna, Luigi, 131
Brenner, Robert, 163, 165
Bresadola, Marco, 159
Briatore, Luigi, 110, 158
Brittuliano, Johannes, 36
Brizzi, Gian Paolo, 59, 73, 78
Brockliss, Laurence W. B., 104
Broman, T., xvii
Brooke, John Hedley, 137
Brou, Alexandre, 137
Brugi, Biagio, 85, 109
Bruno da Longoburgo, 65
Bruno, Giordano, 3, 11
Bucciantini, Massimo, 13, 36, 61
Buffon, Georges, 122

Cabeo, Nicolas, 95, 107
Caetani, Luigi, 24
Caffiero, Marina, 160
Caizzi, Bruno, 14, 57
Caldani, Leopoldo Marcantonio, 77, 150
Calliachi, Nicolo, 70
Calogerà, Angelo, 112, 130
Camerarius, Joachim, 22, 26, 36, 37
Campanella, Tommaso, 11, 22-23, 31, 33-35, 39-40
Canone, Eugenio, 35, 40
Canterzani, Sebastiano, 150-51
Capp, Bernard, 40
Capponi, Luigi, 24
Capra, Carlo, 160
Capuana, Fabrizia, 159
Capucci, Giambattista, 42, 52, 61
Capucci, Martino, 129, 160
Caputo, Cosimo, 37
Caracciolo, Alberto, 81
Caravale, Mario, 81
Carburi, Marco, 77
Cardano, Girolamo, 4, 21, 24, 25, 34, 38
Cardwell, D. S. L., 165
Caro, Francesco, 95, 107
Carradori, Gioacchino, 152, 160
Carranza, Nicola, 84, 85
Carré, Louis, 98, 109

Carrithers, David, 157
Carugo, Adriano, 83
Carvalho e Melo, Sebastião João de, 127
Cascio Pratilli, Giovanni, 80-82
Casini, Paolo, xvi, 130, 131, 132
Cassini, Giovanni Domenico, 43, 49, 50, 60, 97
Cassini, Jacques, 123
Castellani, Carlo, 158
Castelli, Benedetto, xii, 64, 83, 121, 134
Castelli, Patrizia, 81
Castiglione, Baldassarre, 5
Castronovo, Valerio, 129, 157
Cavalieri, Bonaventura, 44, 128
Cavallari-Murat, Augusto, 107
Cavazza, Marta, 85, 134, 159
Celsus, Aulus Cornelius, 122, 134
Cerati, Gaspare, 84
Cesi, Federico, 6, 15, 63, 75, 78, 82, 87, 102, 103
Cestoni, Giacinto, 52, 61
Chartier, Roger, 84
Chautard Du Clos, Giuseppe Antonio, 120
Chechel, Gasparo, 54, 61
Chouet, Jean, 89
Christine of Lorraine, 9
Ciampini, Giovanni Giusto, 45, 53, 69, 82
Ciancio, Luca, 158, 160
Cioli, Andrea, 18
Cipolla, Carlo, 83
Ciriacono, Salvatore, 155, 164, 165
Ciuffoletti, Zeffiro, 155, 157
Clark, William, 155
Clarke, Samuel, 109
Clarkson, Leslie, 37
Clavius, Christopher, 26
Clement XIII (Carlo Rezzonico), 127
Cochrane, Eric, 80, 82, 85, 103, 130, 154, 156
Codogno, Ottavio, 6
Cohen, H. Floris, xvii
Cohen, I. Bernard, 109, 110, 160
Coing, Helmut, 78
Coleman, William Oliver, 155
Coletti, Francesco, 154
Collina, Maria D., 129
Colonna, Anna, 33

Commandino, Federico, 36
Concina, Daniele, 114, 126
Condamine, Charles Marie de la, 141, 156
Condillac, Etienne Bonnot de, 77
Condorcet, Jean-Antoine-Nicolas Caritat, marquis de, 150
Contardi, Simone, 158
Conte, Emanuele, 78
Conti, Antonio, 96, 107
Contin, Tommaso Antonio, 127
Contini, Alessandra, 155
Copenhaver, Brian P., 35
Copernicus, Nicholas, 1, 2, 4, 9, 11, 14, 22, 36, 44, 128
Coppa, Emilio, 139
Cordero di Montezemolo, Virginia, 80
Cornacchini, Pietro, 125, 136
Cornelio, Tommaso, 52, 75
Cortese, Nino, 81, 82
Cossali, Pietro, 83, 107
Covi, Lorenzo, 117
Cozzi, Bruno, 86
Creighton, Edward, 38
Cremante, Renzo, 80, 129, 130, 160
Cremonini, Cesare, 106
Crespi, Domenico, 120
Crisciani, Chiara, 85
Crosland, Maurice, 79, 103
Crucitti Ullrich, Francesca Bianca, 133
Cucchi, Giovanni Antonio, 90, 104
Cursay, Marquis de, 120
Custodi, Pietro, 157

D'Addio, Mario, 13, 155
D'Alessandro, Alessandro, 81
Dallari, Umberto, 72, 81
Darnton, Robert, 160
Dati, Carlo, 44
Daumas, Maurice, 110
Davi, Maria Rosa, 79
De Angeli, Elvezia, 134
De Benedictis, Giovanni Battista, 113
De Bernardin, Sandro, 79, 81, 82
De Frede, Carlo, 81
De Graff, Regnier, 94
De Maddalena, Aldo, 85
De Mas, Enrico, 37
De Robertis, G., 15
De Rosa, Stefano, 80, 81, 84
De Vivo, Francesco, 107

De Waard, Cornélius, 18, 58
De Zan, Mauro, 130
Dear, Peter, 15
Debus, Allen G., 35, 39
Dechales, Claude François, 97
Del Fante, Alessandro, 15
Del Gratta, Rodolfo, 80
Del Negro, Piero, 86, 132
Del Papa, Giuseppe, 54, 62
Deleule, Didier, 38
Della Porta, Giambattista, 15, 37, 122
Delumeau, Jean, 83, 131
Democritus, 52
Desaguliers, Jean Théophile, 102
Descartes, René, 42, 51, 55, 88, 91, 95, 99, 123, 125, 126, 128
Desroussilles, François Dupuigrenet, 80-82
Devaux, Jean, 62
Di Capua, Lionardo, 51-52, 61, 122
Di Pietro, Pericle, 158
Dibon, Paul, 58
Dimsdale, Thomas, 143
Dini, Alessandro, 57
Diodati, Elio, 16
Di Simone, Maria Rosa, 79
Dipper, Christof, 86
Divini, Eustachio, 49
Dollo, Corrado, 59, 79
Dolman, Claude E., 158
Donati, Claudio, 131
Donati, Vitalino, 94
Doni Garfagnini, Manuela, 58
Dooley, Brendan, 35, 61, 79, 85, 107, 111, 132, 154
Drake, Stillman, 13, 16, 134
Du Hamel, Jean, 89
Duchesneau, François, 158
Dykmans, M., 36

Eamon, William, 13, 59
Edquist, Charles, xvi
Eisenstein, Elizabeth, xvii
Ermini, Giuseppe, 81
Ernst, Germana, 35, 37, 39, 40
Eschinardi, Francesco, 45, 59
Este, Francesco III d', 127

Fabbri, Gaetano, 126
Fabbroni, Giuseppe, 77
Fabri de Peiresc, Nicolas Claude, 6, 43

Fabri, Agostino, 60
Fabricio d'Acquapendente, Girolamo, 71
Fabroni, Angelo, 57, 81
Facciolati, Jacopo, 110
Fadda, Bianca, 156
Falco, Giorgio, 156
Fantazzini, P., 107
Fantoni, Filippo, 4
Fara, Patricia, 160
Fardella, Michelangelo, 51, 55, 84
Farley, John, 158
Fasano Guarini, Elena, 80
Favaro, Antonio, 15, 35, 83, 103
Feldhay, Rivka, 13
Felice, Costanzo, 14
Femiano, Salvatore, 84
Ferchault de Réaumur, René-Antoine, 140
Ferrari, Giorgio E., 60
Ferrari, Marco, 61, 134
Ferro, Scipione dal, 4
Ferrone, Vincenzo, xvii, 109, 130, 134-37, 158, 165
Ferroni, Giuseppe, 47
Festa, Egidio, 13
Field, J. V., 39
Filangieri, Gaetano, 145, 151
Filicaia, Bartolomeo, 32
Findlen, Paula, 159
Finocchiaro, Maurice, 16
Finzi, Roberto, 156
Fior, Antonio Maria, 4
Fiorani, Luigi, 35
Fioravanti, Leonardo, 4, 14, 50
Firmian, Carlo, 147
Firpo, Luigi, 37
Fisch, Max, 85
Fitzgibbons, Athol, 155
Flamsteed, John, 53, 96
Florio, Bernardo, 91
Fludd, Robert, 27, 39
Fontana, Felice, 77, 150
Forti, Fiorenzo, 156
Fortis, Alberto, xv, 144-48
Foscarini, Antonio, 11
Fowler, Richard, 160
Fox, Christopher, 157
Fracastoro, Girolamo, 90, 104
Fragnito, Gigliola, 18
Franchi, Giansebastiano, 121

Franklin, Benjamin, 120, 146
Frisi, Paolo, 85, 128
Frova, Giuseppe, 126
Fuchs, Leonard, 4, 14
Fuller, Steve, xvii
Fumi, Giampiero, 156

Gabrieli, G., 15
Gabrielli, Pirro Maria, 75
Galen (Claudius Galenus), 6, 26, 64-65, 90, 93, 104-6,
Galiani, Ferdinando, 139
Galilei, Galileo, xi, xiii, xv, 1, 20-21, 30, 33, 35, 36, 39, 40, 41, 64, 70, 88, 93, 95, 98, 103, 113, 121-22, 128, 162
Galluzzi, Paolo, 13, 35, 39, 57, 58, 59, 60, 61, 84, 85, 103, 134
Galvani, Luigi, xii, xiii, xvi, 77, 138, 144, 147, 150-51, 153, 159
Gambarin, Giovanni, 18
Garcaeus, Johannes, 24, 38
Gardair, Jean-Michel, 16, 60
Gardi, Andrea, 86
Garibotto, Celestino, 129
Garin, Eugenio, 13, 37, 61, 134
Garofalo, Silvano, 158
Garofano, Ivan, 105
Garvey, William D., xvii, 60
Gasparini, Carolina, 129
Gassendi, Pierre, 3, 42, 44, 46, 56, 95, 98, 108, 128
Gatto, Romano, 59
Gaude, Francisco, 40
Gaurico, Luca, 21, 24, 30, 38
Generali, Dario, 104
Geneva, Ann, 35, 40
Gennari, Giuseppe, 107
Genorini, Michele, 122
Genovesi, Antonio, 125, 136, 139, 140
Gerdil, Sigismondo, 77
Gesner, Konrad, 47
Geymonat, Ludovico, 13, 103
Gherardi, Gherardo, 22
Ghisleri, Francesco Maria, 24
Giannone, Pietro, 113, 130
Giardina, Leonardo, 159
Giarrizzo, Giuseppe, 157
Giattini, Giovanni Battista, 47, 59
Gibba, Alessandro, 79
Gigli, Giacinto, 19, 35

Gillispie, Charles Coulston, 135, 155, 165
Gingerich, Owen, 16, 17, 36
Gini, Corrado, 86
Ginzburg, Carlo, 84
Giordano, Davide, 79
Giuliano, Giambattista di, 26
Gliozzi, Mario, 108
Goclenius, Rudolph, 22, 24, 36
Goethe, Johann Wolfgang von, 153
Gogava, Antonius, 26, 29, 36, 38
Goidanich, Gabriele, 158
Golinski, Jan, 155
Gómez López, Susana, 61
Gonzaga, Ferdinando I, 81
Gori, Giambattista, 109, 132
Grab, Alexander I., 154-55
Grafton, Anthony, 38, 39, 79, 84
Grandi, Jacopo, 91
Granvelle, Nicolas Perrenot de, 83
Grassi, Orazio, 8
Grassi, Orazio, 8
Gravesande, Willem Jakob 's, 99, 100, 102, 109-10, 124, 146
Gregory, Tullio, 108
Gremigni, Elena, 164
Grendler, Paul, 15, 81
Grenier, Jean-Yves, 35
Griffini Rosnati, Grazia Maria, 164
Grimaldi, Costantino, 122, 134
Griselini, Francesco, 143, 157
Gronda, Giovanna, 129
Gross, Hanns, 131, 136, 155
Gross, Paul R., xvii
Gualazzini, Ugo, 80
Gualdo, Paolo, 17
Guderzo, Giulio, 85
Guericke, Otto von, 98
Guerlac, Henry, 109, 135
Guerrini, Giuseppe, 157
Guerrini, Pietro, 162, 164
Guglielmini, Domenico, 60, 99, 109, 121, 144
Gullino, Giuseppe, 106, 107
Gusdorf, Georges, 157

Habermas, Jürgen, xvii
Hacking, Ian, xvii
Hackmann, Willem Dirk, 111
Hale, J. R., 82
Hall, Marie Boas, 61

Hall, Rupert, 57
Hall, Thomas S., 158
Haller, Albrecht von, 71, 146, 150
Hallyn, Fernand, 39
Hankins, Thomas L., 133
Harriot, Thomas, 30
Harvey, William, 88, 90, 91, 105
Hauksbee, Francis, 99, 109
Heilbron, John L., 110, 158-59
Heinrich, Johann Jacob, 53
Helmont, Jean Baptist van, 65, 91
Hermann, Jacob, 70
Herriott, Thomas, 6
Hevelius, Johannes, 108
Heyd, Michael, 104
Hilfstein, E., 13
Hipparchus, 22
Hippocrates, 64-65, 91
Hodierna, Giambattista, 23, 122
Hooke, Robert, 98
Horn, Georg, 97
Houte, Gillis van den, 162
Hübner, Wolfgang, 39
Huffman, William H., 39
Huguetan, Jean, 43, 58
Hume, Joseph, 151
Huygens, Christiaan, 43, 53, 62, 128, 162

Imberciadori, Ildebrando, 155
Inchofer, Melchiorre, 18
Infelise, Mario, xvii, 18, 129-30, 133, 137, 157
Intieri, Bartolomeo, 76
Ioli, Antonio, 83

Jansen, Cornelius, 77
Jardine, Lisa, 79, 84
João I, 127
Jona, Alfredo, 159
Joseph II, 150, 151
Julia, Dominique, 84

Kagan, Richard L., 84
Katzen, May, 60
Kaunitz, Wenzel Anton von, 77
Keil, John, 125
Kepler, Johannes, 6, 22, 24, 27, 34, 36, 39, 123
Kircher, Athanasius, 95, 122
Kirshner, Julius, 80

Klibansky, Raymond, 39
Koyré, Alexandre, 13, 109, 110
Kristeller, Paul Oskar, 81, 103
Kronick, David A., 60

La Chalotais, Louis-René de Caradeuc de, 137
Lagrange, Louis, 78, 162
Lami, Giovanni, 112, 114, 116, 126, 130, 140
Lamponi, Francesco, 30, 32
Lanaro, M., 60
Landes, David S., 133, 164-65
Lapi, Giovanni, 139
Lasswitz, Kurd, 108
Lattis, James, 38
Lauria, Donatella, 61
Lazzari, Pietro, 116
Le Boë (Franciscus Sylvius), Franz de, 105
Lecce, Michele, 154, 156
Lecchi, Antonio, 115
Leeuwenhoek, Antonie van, 94, 106, 121
Leibniz, Gottfried Willem, 109, 118
Leowitz, Cyprian, 22, 23, 37
Lerner, Michel-Pierre, 38
Lesnodorski, Boguslaw, 78
Leti, Gregorio, 41, 43, 57, 162, 164
Levitt, Norman, xvii
Lewis, Christopher, 79
Liberi, Pietro, 54
Liguori, Alfonso Maria de', 131
Lindberg, David C., 13, 35
Lindhout, Henricus, 22, 36
Litchfield, R. Burr, 80
Locke, John, 77, 151
Logan, Gabriella Berti, 159
Lombardini, Elia, 83
Lorgna, Anton Maria, 162
Lubienski, Stanislas, 97
Lucretius, 54
Lupano, Alberto, 81

Mabillon, Jean, 61, 116
Macchi, Giovan Battista, 126, 136
Maccope, Alessandro Knips, 65, 79
Macfarlane, Alan, 163, 165
Maffei, Scipione, 75, 99, 102, 107, 109, 113, 120, 124, 126, 129, 133, 137
Maffei, Tommaso Pio, 95

Maffiodo, Barbara, 157
Maffioli, Cesare S., xvi
Magalotti, Lorenzo, 43, 50, 71, 84, 121, 125
Magini, Giovanni Antonio, 7, 36, 39
Magiotti, Raffaele, 58
Magliabechi, Antonio, 43, 103
Magnani, Ippolito, 55
Magnus, Albertus, 153
Maio, Romeo de, 35
Maiocchi, Roberto, 59
Malagola, Carlo, 80
Malebranche, Nicolas, 98
Malespina, Vitellio, 32
Malpighi Marcello, 41, 44, 49, 52, 53, 57, 61, 64, 65, 71, 79, 88, 90, 91, 121, 144
Mandelbaum, Jonathan, 160
Manetti, Saverio, 140, 143
Manfredi, Eustachio, 123, 130
Mango Tomei, Elsa, 79
Manzini, Paola, 159
Marcelli, Ugo, 86
Marchetti, Alessandro, 44, 55, 84
Marcialis, Maria Teresa, 136
Marcocci, Massimo, 137
Marini, Ottavio, 25
Marini, Paola, 137
Marrara, Danilo, 80, 83, 84
Marsili, Anton Felice, 75, 84
Marsili, Luigi Ferdinando, 49, 75
Martelli, Francesco, 164
Martens, Martinus, 101
Martinelli, Christian, 97
Martinovic, Ivica, 132
Marzari, Antonio, 14
Masat Lucchetta, Paola, 105
Maternus, Julius Firmicus, 22, 36
Maupertuis, Pierre-Louis, 123
Maurolico, Francesco, 4, 81
Mayer, Simon, 7, 8
Maylender, Michele, 85
Mazzei, Filippo, 160
Mazzetti, Serafino, 104
Mazzolini, Renato G., 158
McClellan, James E., III, 154, 165
Meadows, A. J., xvii, 60
Medici, Caterina de', 81
Medici, Cosimo I de', 66, 68, 80, 82, 83
Medici, Cosimo III de', 74, 162, 164

Medici, Giovanni de', 33
Medici, Giuliano de', 16
Medici, Leopoldo de', 45, 84
Melanchthon, Philip, 22, 26, 37, 38
Meli, Domenico Bertoloni, 57
Melli, Sebastiano, 75
Memmo, Andrea, 138
Mendelsohn, Everett, 62
Mengoli, Pietro, 122, 134
Menzini, Benedetto, 114
Merkel, Ingrid, 35
Mersenne, Marin, 6, 42, 43
Mesmer, Franz, xvi, 146, 153
Micheli, Gianni, 134
Micheli, Pier Antonio, 121, 144
Michelotti, Francesco Domenico, 78
Middleton, William E. Knowles, 59, 82, 84, 108
Migliorini, Anna Vittoria, 130
Milani, R., 158
Milton, John, 12
Mirri, Mario, 155
Mirto, Alfonso, 59
Moioli, Angelo, 157
Molinetto, Michelangelo, 70
Momigliano, Arnaldo, 61
Montalenti, Giuseppe, 15, 159
Montanari, Geminiano, 48, 49, 50, 55, 57, 60, 62, 88, 89, 96, 103
Montelatici, Ubaldo, 138, 140
Montesquieu, Charles Louis de Secondat, baron de, 151
Moran, Bruce T., xvi
Morandi, Orazio, xiv, 20, 41
Moray, Robert, 57
Moreland, Samuel, 49
Morgagni, Giambattista, 65, 70, 79, 107, 113, 122, 134, 146, 148
Morpurgo, Edgardo, 83
Moscheo, Rosario, 13, 81
Moschini, Giannantonio, 107
Moss, Jean Dietz, 16, 58
Mozzarelli, Cesare, 86
Mucillo, M., 59
Muratori, Ludovico Antonio, 106, 114, 130, 132, 140, 156
Musschenbroek, Jan, 110
Musschenbroek, Petrus van, 99, 100, 102, 124, 141, 146
Musschenbroek, Samuel van, 110
Musson, Albert Edward, 110

174 Index

Muti, Tiberio, 24
Naibod, Valentin, 36
Napier, John, 6
Nardi, Antonio, 58
Nardo, Dante, 107, 134
Naux, Charles, 16
Nazari, Francesco, 42
Needham, John Turberville, 147
Neri, Pompeo, 139
Newcomen, Thomas, 163
Newton, Isaac, 70, 83, 107, 109, 110, 115, 123, 126, 128, 135, 146
Niccolini, Francesco, 18
Niceron, Jean-Pierre, 104
Nichetti Spanio, M., 106
Nifo, Agostino, 27, 38
Nigrisoli, Francesco Maria, 122
Nollet, Jean-Antoine, 101, 146, 149
Nonni, Giorgio, 14
North, John David, 38
Novarese, Daniela, 81, 82
Nussdorfer, Laurie, 37

Odoardi, G., 38
Offusius, Jofrancus, 27, 39
Okruhlik, Kathleen, 133
Oldenburg, Henry, 42, 43, 49
Olivieri, Luigi, 35
Olmi, Giuseppe, 16, 82
Oporinus, Joannes, 38
Oppenheimer, J. Robert, 3
Orlandi, Giuseppe, 156
Osbat, Luciano, 57, 84, 130
Osiander, Andreas, 44

Paciaudi, Paolo Maria, 113
Pagel, Walter, 105
Paitoni, Jacopo Maria, 129
Palcani, Luigi, 131
Pallavicino, Sforza, 9, 11, 17, 58
Palm, L. C., 104
Palmer, Robert, 80
Pancaldi, G., 159
Pancino, Maria, 85, 110
Pane, Luigi dal, 157
Panofsky, Erwin, 39
Paolella, Alfredo, 37
Paoli, Germano, 131
Papparelli, Gioacchino, 37

Paracelsus (Philippus Theophrastus Bombastus von Hohenheim, called, 51, 65, 71, 153
Parri, Maria Grazia, 155
Parsons, Talcott, 165
Partner, Peter, 37
Pascal, Blaise, 98, 137
Pascoli, Alessandro, 56, 62
Pasqualigo, Zacharia, 18
Pasta, Renato, 155-57
Pastor, Ludwig von, 137
Pastore Stocchi, Manlio, 83, 154
Patuzzi, Gian Vincenzo, 136
Paul V (Camillo Borghese), 8
Pavone, Mario, 37
Pazzini, Adalberto, 79
Pedersen, Olaf, 78
Pemberton, Henry, 124, 135
Pepe, Luigi, 19, 82
Pera, Marcello, 159
Peroni, Baldo, 86
Petrocchi, Giorgio, 131
Piaia, Gregorio, 85
Piana, Celestino, 83
Piccolini, Marco, 159
Piccolomini, Alessandro, 5, 12, 15, 18
Pico della Mirandola, Giovanni, 24
Pietro Leopoldo, 140
Pighetti, Clelia, 107
Pignatelli, Giuseppe, 130
Pii, Eluggero, 86, 156
Pilati, Carlo Antonio, 138, 142, 154
Pini, Antonio Ivan, 73
Pistorini, Giacomo, 86
Pitt, Joseph C., 133
Pius IV (Giovanni de' Medici), 68
Piva, Franco, 165
Pivati, Giovanni Francesco, 101, 146, 158
Plancus, Janus, 113
Plata, Francesco Maria, 115
Plato, 6
Poleni, Giovanni, xv, 70, 83, 88, 94, 96-97, 99, 101-2, 107, 144
Poletti, Orazio, 120
Pomata, Gianna, 37
Pomponazzi, Pietro, 64, 79
Porcia, Giovan Artico di, 104-6
Porta, Pier Luigi, 157
Porter, Roy, 37, 78, 157
Porzio, Lucantonio, 42

Poupard, Paul, 13
Preto, Paolo, 154
Prodi, Paolo, 78
Prontera, Gemma, 156
Ptolemy, 6, 22, 25, 26, 27, 32, 36, 39
Pugliese, Patri Jones, 136
Puliatti, Pietro, 19
Pythagoras, 106

Querini, Andrea, 139
Querini, Angelo Maria, 116

Rachum, Ilan, 37
Radder, H., xvi
Raet, Willem de, 162
Raimondi, Ezio, 16, 78, 85, 103
Ramazzotti, S., 110, 158
Ramelli, Agostino, 163
Rantzau, Henrik, 22, 36
Raspadori, Francesco, 78
Ray, John, 43
Raynal, 150
Redi, Francesco, 49, 56, 62, 121
Redondi, Pietro, 13
Regiomontanus, Joannes, 25, 26, 36
Régis, Pierre Sylvan, 98, 108
Reif, Patricia, 59
Reinbold, Anne, 15,
Reinhold, Erasmus, 24, 36
Remondini, Giambattista, 120
Remondini, Giovanni Antonio, 120
Revel, Jacques, 84
Riccati, Vincenzo, 113, 115, 118
Ricci, Lorenzo, 127
Ricci, Michelangelo, xii, 41
Riccio, Pier Francesco, 82
Riccioli, Giovanni Battista, 44, 47, 59, 84, 96, 122
Ricciotti, Gino, 19
Riccomanni, Luigi, 142
Ricuperati, Giuseppe, 60, 79, 85, 129, 157
Ridder-Symoens, Hilde de, 78
Ridolfi, Ludovico, 33
Ridolfi, Nicolo, 24
Ridolfi, Vincenzo, 82
Rigotti, Francesco, 160
Ristori, Giovanni, 150-51
Rivière, Lazare, 104
Rizzetti, Giovanni, 124
Robbins, F. E., 40

Roberval, Gilles Personne de, 3
Robinet, André, 109
Robinson, Eric, 110
Robortello, Francesco, 67
Rodolico, Francesco, 134
Roe, Shirley. A., 159
Roggero, Marina, 78
Roma, Giuseppe, 75
Romagnoli, Sergio, 129, 130
Rombai, Leonardo, 131, 155, 157
Rondinelli, Simon Carlo, 33
Rosa, Enrico, 130, 137
Rosa, Mario, 36, 129, 131, 132
Rose, Paul Lawrence, 13
Rosino, Leonida, 13
Rossello, Timoteo, 47
Rossetti, Donato, 12, 41, 48, 51, 59, 122
Rossetti, Lucia, 104
Rossi, Paolo, 38, 159, 160
Rossi, Pietro, 36
Rotelli, Ettore, 85
Rother, Wolfgang, 104
Rotondi, Clementina, 155
Rotta, Sergio, 62, 85, 103, 104, 108
Rousseau, Jean-Jacques, 151
Roveda, Valentino, 120
Rowbottom, Margaret, 111
Rozzo, U., 36
Rubbi, Andrea, 128
Ruestow, Edward G., 104
Russell, Andrew W., 80
Russo, François, xvii
Ruzzante, Angelo Beolco, called, 17

Sabia, Mario, 105
Saccenti, Mario, 58
Sacco, Pompeo, 91, 99, 109
Sacrobosco, Giovanni, 65
Saibante, Mario, 74, 86
Sala, Torello, 35
Saladini, G. A., 110
Salandin, Gian Antonio, 85
Salomoni, Pier Maria, 124
Salvemini, Biagio, 35
Sandri, Giacomo, 75
Sangro, Raimondo di, principe di S. Severo, 119, 133
Santillana, Giorgio de, 3, 13
Santillo, Marco, 155
Santorio, Santorio, 98, 105

Santucci, Simonetta, 129
Sarpi, Paolo, 10
Savelli, Roberto, 106
Savoy, Victor Amadeus II, duke of, 76
Saxl, Fritz, 39
Sbaraglia, Giovanni Girolamo, xv, 88, 89, 104, 105
Scaglia, Desiderio, 24, 32
Scarpa, Antonio, 77
Schaffer, Simon, 155
Scheiner, Christopher, 7
Scheuchzer, Johann Jacob, 70
Schipa, Michelangelo, 81
Schirrmacher, Thomas, 13
Schmitt, Charles B., 13, 38, 79, 81, 103
Schullian, Dorothy M., 60
Scilla, Agostino, 144
Scioscioli, Donato, 130-33
Scottoni, Giovanni Francesco, 143, 157
Segneri, Paolo, 114
Segre, Michael, xvi
Sellari, Bortolo, 87, 103
Senarega, Stefano, 22
Sereni, Emilio, 154
Serenus Sammonicus, Quintus, 122, 134
Serpetro, Giovanni, 47, 59
Serra, Armando, 15
Serra, Salvatore, 49, 55, 60
Serristori, Antonio, 139
Servetus, Michael, 91, 105
Sforza, Francesco, 82
Sguario, Eusebio, 144, 146
Shapin, Steven, 78
Shattuck, Roger, xvii
Shea, William R., 13
Shelley, Mary, 153, 160
Shumaker, Wayne, 35
Signorotto, Gianvittorio, 137
Sigonio, Carlo, 67
Silvestri, Daniela, 158
Simionato, Giustina, 79
Simon, Gérard, 39
Simoncelli, Paolo, 13
Siraisi, Nancy G., 38, 79
Sirleto, Card. Guglielmo, 14
Sixtus V (Felice Peretti), 22
Sloane, Hans, 70
Smith, Joseph, 107
Solani, Bartolomeo, 120
Sommervogel, Carlos, 130, 131

Soppelsa, Maria Laura, 83, 85, 103
Sordi, Bernardo, 155
Sottili, Agostino, 82
Spadon, Giovanni Maria, 54
Spallanzani, Lazzaro, xii, xv, 77, 138, 144, 146, 147, 148, 153
Spampanato, Vincenzo, 35
Spano, Nicola, 81
Speroni, Sperone, 6, 15
Stabile, Giorgio, 103
Stano, Gaetano, 38
Stella, Pietro, 131
Steno, Nicolas, 43
Stephenson, Bruce, 39
Stewart, Hugh Fraser, 137
Stewart, Larry, 155
Stoeffler, Johann, 36
Stone, Lawrence, 83
Stromholm, P., 108
Sutton, Geoffrey V., 155, 160

Tacquet, André, 97, 108
Taddei, Giovanni, 54
Tagliaferri, Amelio, 106
Tannery, Paul, 18, 58
Tanucci, Bernardo, 76, 126, 130, 139, 147
Targioni Tozzetti, Giovanni, xv, 113, 122, 140-41, 144, 148, 156-57
Tartaglia, Niccolò, 4, 14, 162
Tartarotti, Girolamo, 125
Tassoni, Alessandro, 12, 19
Tassot, Dominique, 13
Tawney, R. H., 163, 165
Tebaldi, Egidio, 26, 38
Tega, Walter, 80, 130
Tenca, Luigi, 108
Tentorio, Marco, 107
Tester, S. J., 35
Testi, Lodovico, 56, 91
Thomas, Keith, 23, 37
Thorndike, Lynn, 39
Thouvenel, Pierre, 147
Tibelli, Giovanni, 53
Toaldo, Giuseppe, 77, 134
Todeschi, Claudio, 139, 143
Toellner, R., 159
Tognarini, Ivan, 160
Toledo, Pedro de, 68, 82
Tomaselli, Sylvana, 157
Tomitano, Bernardino, 64

Tondo, Luigi, 81
Toniolo Fascione, Maria Claudia, 81
Torcellan, Gianfranco, 157
Torelli, Lelio, 67, 68
Torricelli, Evangelista, xii, 44, 47, 53, 163
Torrini, Maurizio, 13, 36, 58, 59, 61, 85, 134
Tortarolo, Edoardo, 160
Tramontin, Silvio, 107
Tranfaglia, Nicola, 60, 157, 129
Trevisan, Bernardo, 96
Trinci, Cosimo, 141
Trionfetti, Giambattista, 52
Trionfetti, Lelio, 91
Tron, Andrea, 138, 139
Troncarelli, Fabio, 35
Truesdell, Clifford, 109
Tucci, Pasquale, 132
Turner, R. Steven, 83

Uguccioni, Giambattista, 139
Underwood, E. Ashworth, 85
Urban VIII (Maffeo Barberini), xiv, 3, 11, 12, 20, 30-34, 40
Usimbardi, Francesco, 22, 24

Vaccari, Ezio, 156, 157, 159
Vaillè, Eugène, 14
Valiano, Sigismondo, 54
Valletta, Giuseppe, 51, 52
Valletta, Nicola, 153, 160
Valli, Eusebio, 160
Vallisneri, Antonio, jr., 145, 158
Vallisneri, Antonio, xv, 52, 55, 61, 88, 89, 91, 92, 93, 102-06, 114, 121, 131, 144, 158
Van den Daele, Wolfgang, 62
Van Helden, Albert, 58
Vandenhout, Egidio, 162
Vanni, Giovanni Francesco, 60
Varni, Angelo, 78
Vasoli, Cesare, 15, 35, 40, 61
Vecchi, Alberto, 131, 133
Vedrine, Helène, 37
Vegetti, Mario, 105
Veggetti, Alba, 86
Venturi, Franco, 85, 130, 136-37, 154-58
Venuti, Ridolfino, 112
Veratti, Giuseppe, 120, 149, 159

Verri, Alessandro, 76
Verri, Pietro, 76, 147, 150, 151
Verrua, P., 107
Vesalius, Andreas, 71
Vettori, Pietro, 36
Vianelli, Giuseppe, 133
Vickers, Brian, 35
Vico, Giambattista, 52
Vieri, Francesco de', 79
Villani, Filippo, 139
Villiers, Christophe, 18
Vinci, Leonardo da, 163
Visceglia, Maria Antonietta, 37
Visconti, Ignazio, 127
Visconti, Raffaele, 32, 33
Visentini, Antonio, 107
Vitruvius, 5
Vivarini, C., 86
Vivian, Frances, 107
Viviani, Vincenzo, 122
Voghera, G., 86
Volpi, Giuliana, 72, 83, 84
Volta, Alessandro, xiii, xvi, 77, 138, 152-53
Voltaire, 143, 148, 151
Voss, Isaac, 97

Walker, D. P., 38
Wallace, William, 79
Waller, William, 53, 61
Wandruszka, Adam, 155
Waquet, Françoise, 58
Waquet, Jean-Claude, 155
Watson, Richard A., 109
Weber, Giorgio, 58
Weber, Max, 163, 165
Weingart, Peter, 62
Westfall, Richard S., 58
Westman, Robert S., 13, 14, 16, 35
Whitley, Richard, 62
Willis, Thomas, 93, 104
Wise, Norton, xvii
Wokler, Robert, 157
Wolff, Christian, 102
Wurstisen, Christian, 89

Ximenes, Lionardo, 115, 131, 139

Zabarella, Giacomo, 26
Zaccaria, Francesco Antonio, xv, 112-31

Zaghi, Carlo, 160
Zambelli, Paola, 35, 37, 84
Zaninelli, S., 156
Zanon, Antonio, 142-43, 157
Zanotti, Francesco Maria, 118
Zardin, Danilo, 36
Zendrini, Bernardino, 96
Zeno, Apostolo, 102, 116, 120, 133
Zorzi, Marino, 36
Zorzoli, Maria Carla, 81
Zugmesser, Johann Entel, 7

About the Author

Brendan Dooley is Research Coordinator of the Medici Archive Project, based in Florence and New York. He taught for many years at Harvard University, and has held fellowships at the American Academy in Rome, the Institute for Advanced Study in Princeton, and the European University Institute. Major publications include *Science, Politics and Society in Eighteenth-Century Italy: The* Giornale de' letterati d'Italia *and its World* (New York: 1991), *Italy in the Baroque: Selected Readings* (New York: 1995), *The Social History of Skepticism: Experience and Doubt in Early Modern Culture* (Baltimore: 1999), *Giovanni Baldinucci: Ricordi. Peste, guerra e società a Firenze nel Seicento* (Florence: 2001) and, with Sabrina Baron, *The Politics of Information in Early Modern Europe* (London: 2001).